T0318398

THE
ORDER
Of
DAYS

THE
ORDER
OF
DAYS

THE MAYA WORLD

AND THE TRUTH ABOUT 2012

DAVID STUART

THREE RIVERS PRESS

New York

For Peter and Richard

Published in the United States by Harmony Books,
an imprint of the Crown Publishing Group,
a division of Random House, Inc., New York.
www.crownpublishing.com

Three Rivers Press and the Tugboat design are
registered trademarks of Random House, Inc.

Originally published in hardcover in the United States by
Harmony Books, an imprint of the Crown Publishing Group,
a division of Random House, Inc., New York, in 2011.

Library of Congress Cataloging-in-Publication Data
Stuart, David, 1965–
The order of days : the Maya world and the truth about 2012 / David Stuart. — 1st ed.
p. cm.
Includes bibliographical references.
1. Maya calendar. 2. Maya astronomy. 3. Maya cosmology. 4. Maya —
Prophecies. 5. End of the world (Astronomy) 6. Two thousand twelve, A.D. I. Title.
F1435.3.C14S78 2011
529'.32978427 — dc22
2010028515

ISBN 978-0-385-52727-9
eISBN 978-0-307-72081-8

Book design by Chris Welch

First Paperback Edition

CONTENTS

AUTHOR'S NOTE
THE SPELLING AND PRONUNCIATION
OF MAYAN WORDS

The spelling of Mayan words seems always to be a complex matter, and readers may well be confused by their look and the seemingly exotic sounds they represent. One important sound feature of Mayan is the glottal stop, represented generally by an apostrophe before a vowel, and sometimes after consonants such as *k*, *t*, or *ch* (hence *k'*, *t'*, or *ch'*). The glottal stop involves an abrupt interruption or obstruction of airflow during speech, and while not a part of standard English phonology, many English speakers do often make use of it, as in the gap between the two utterances of *uh oh*. In Mayan these are often found after consonants, as in *ch'ak*, or *bak'*. It's important to stress that Mayan languages always distinquish words with glottalized and unglottalized consonants, so that in Yukatek Mayan, for example, *kan* means "snake," but *k'an* means "yellow." The vowels of Mayan are much like those of Spanish.

In this book I have generally followed the standard spelling conventions used by specialists in Maya studies, with some important exceptions. The proper names of time periods, days and months, for example, conform to long-standing conventions and may not always agree with the preferred technical orthography used by modern linguists. *Ajaw* is the word for "king" in classic Mayan, but the similar day name I prefer to spell it as *Ahaw*, largely because the Yukatek Mayan language of the sixteenth century, where the day names we use come from, did not have a *j*. There are similar lines of reasoning behind some of the spellings that might at first look inconsistent or unconventional.

Some readers will no doubt wonder about the different usage of "Maya" and "Mayan." The words are not synonymous: "Maya" is a general ethnic term and adjective ("the ancient Maya" or "Maya art") whereas "Mayan" is reserved for the language group ("Mayan languages," "proto-Mayan," etc.). This distinction follows the general conventions used in recent years by anthropologists, linguists, and archaeologists. The plural forms "Mayas" or "Mayans" are sometimes used in popular writings (less so in academic literature) but will be avoided here in preference for the more simple "Maya."

Thirty or so Mayan languages are spoken today in parts of Mexico, Belize, and Guatemala, and even in the streets and farm fields of the United States. Each language (Yukatek, K'iche', Ch'ol, among many others) has its own distinctive sound-system and grammatical structure. The language of the ancient hieroglyphic texts I prefer to call "Classic Mayan" is ancestral to only a handful of these modern languages.

PREFACE

In the last half-century, modern scholars have made an astounding intellectual journey. Beginning in the 1950s and '60s, archaeologists and historians finally began the rigorous process of understanding many key aspects of ancient Maya civilization, much of it by "cracking the code" of the elaborate Maya hieroglyphic script, left to us on hundreds of stone monuments and ceramic vessels. This work has enveloped me for most of my life, with my interest in the Maya beginning when I was a boy accompanying my parents on their expeditions to remote jungle ruins back in the 1970s. Over the years I've been incredibly fortunate to participate in this transformation of knowledge, working with many colleagues in diverse fields to bring the ancient Maya from the realm of prehistory into that of history. Now, after several exciting decades, their written record is mostly understood, and it has forever changed our view of Maya history, religion, and culture. I like to think that we're now at a place in the study of the ancient Maya not unlike where Egyptologists were in the early nineteenth century, at which time an ancient civilization suddenly was ripe for study in

grand detail, right on the heels of the decipherment of Egyptian hieroglyphs by Jean-François Champollion. With regard to the Maya, we
are nowadays in a similar "heady" time, albeit with far more scientific
methodology and context in hand than early Egyptology ever had.

But has the popular understanding of ancient Mesoamerica and the
Maya advanced so much? I have to wonder. Lately I find myself confused and even frustrated by what I see in the popular representation
of the Maya in today's media, whether it be in print or on-screen. Seldom can I roam through a large retail bookstore, sit in front of a television, or surf the Web without seeing some reference to the year 2012,
now just a couple of years away as I write this. Many of the books on
2012 have evocative, even alarming titles, such as *Apocalypse 2012: An
Investigation into Civilization's End*; or *The World Cataclysm in 2012: The
Maya Countdown to the End of Our World*; or *2012: The Return of Quetzalcoatl*; or *Maya Cosmogenesis 2012: The True Meaning of the Maya Calendar
End-Date*. According to many of these strange-sounding books and TV
shows—and none of them is ever consistent in its message—the ancient
Maya, having some keen insight into the mystical workings of our
planet and the cosmos, were able to predict that the world would end or
in some way be radically transformed in the year 2012—on the winter
solstice December 21, to be exact (although, again, some sources differ
about the precise day).

This is all complete nonsense. As someone who has studied the
Maya for nearly all of his life, and who specializes in reading their
ancient texts in order to understand their history, religion, and culture,
I have to lay down the line and assert that any such statements about
the Maya predicting the world's demise or, alternatively, some "transformation of consciousness" in 2012 is, to put it as simply and directly
as possible, wrong. Not only wrong, but misleading.

There's something larger at work here, more than just the ideas of
a few kooks who have little interest in real Maya history and culture.
The 2012 hubbub seems to be the most recent in a long chain of related
theories and ideas about Mesoamericans, and the Maya in particular,
depicting them as somehow oddball, not "of this world," or having some
strong mystical link to other realms and dimensions. As early as the

nineteenth century, the emerging accounts of ancient civilizations in Mexico were widely seen as too impossibly advanced to be the handiwork of "Indians," especially among many in the young United States, where native populations were being slaughtered, displaced, and culturally marginalized along an ever-increasing frontier. Some people claimed that the impressive ruins of Central America had to be the works of Phoenicians, Israelites, Scandinavians (?!), or even inhabitants of the lost continent of Atlantis. How could "Indians" have built such great cities and created such artwork?—or so the thinking went. This notion was largely dispelled among scholars by the mid-1800s, as archaeology blossomed and exploration established no doubt about historical and cultural links between the ruined cities and the native inhabitants of the area. But by the twentieth century, the same vein of thinking had morphed somewhat, now depicting the ancient Maya not so much as Old World seafarers but as peaceful, star-gazing intellectuals little concerned with the real world of human affairs. No wonder, perhaps, that by the 1970s, pop culture references to the exotic Maya had them making direct contact with aliens, who, after all, must have built their cities using spaceships. Even today we see the same motif in movies and books. The most recent Indiana Jones film—always good PR for archaeology—shows Maya pyramids in the Amazon, of all places, guarding crystal skulls and flying saucers.

At present, with the ominous year 2012 fast approaching, it isn't surprising that many of these same ideas are back in the news. According to many breathless writers and to TV accounts, the "Maya calendar will end," and the Maya were somehow able to tap in to their mystical knowledge to predict the future of our time. At the very least, through some supposed special connections to the outer cosmos or to the inner being of humanity (depending on what you read), the Maya alone were able to predict a coming end of times, a galactic alignment with the rising sun, and the transformation of human consciousness, among other weird and great insights.

I'm convinced that these outlandish claims about 2012 and its meaning (as if it needs a meaning) are the latest manifestations of a deep-seated idea within the American popular imagination, that Mesoamerican peoples and the "mysterious Maya" are exotic and even, in some way, alien. Fundamentally, this speaks to an inability in much of our own culture, seldom addressed or even recognized, to confront the fact that many ancient Americans were as civilized as their counterparts in the Old World. There's a lot of intellectual baggage behind these thoughts and denials, some going back to centuries-old notions about "the Indian" as a noble savage, or maybe just a savage. Couple these ideas with the profound cultural barrier that exists between the United States and the history and culture of Mexico and Central America, and you have a recipe for major cultural misunderstanding. To me, the whole situation reveals the fact that modern America still has a difficult time grasping the reality of ancient advanced cultures south of our border. If one looks at attitudes about Mexico and illegal "aliens" today, maybe this is not so surprising.

The irony is acute. For just at the moment when decades of hard work in the field and in the libraries have paid off, when we can proudly claim to have cracked the Maya code, most of the loudest "expert" voices are those who simply misrepresent the truth about the Maya, a remarkable people who deserve far, far better. In other words, nearly all of the books and television shows on 2012 are by gurus and spiritualists who wouldn't know a Maya glyph if one hit them on the nose. I'm convinced that the emerging 2012 phenomenon says far more about today's culture and its larger concerns than about the ancient Maya.

This profound misrepresentation of the Maya motivated the writing of this book. In my outlook, the reality of Maya accomplishments is far more interesting and awe-inspiring than the ubiquitous false claims about their culture and what, if anything, they had to say about the future. This book is not about New Age thinking but instead about old age thought, based on what ancient Maya records, as best we can understand them, actually had to say about time and the workings of the cosmos, and about their place—not ours—within this complex worldview. Because 2012 has no prominent role in anything the ancient Maya ever wrote (only one clear reference to the date exists among thousands of inscriptions), this book takes a wider look at the Maya concepts of time and at the philosophy that gave rise to them, at least as we can best understand the two. One of my overarching goals is to provide ample evidence that the real intellectual achievements of ancient Maya timekeeping and worldview—incomplete as what has been passed down to us is—are far more impressive and remarkable than any of the more popular outrageous and wrongheaded claims about Maya uniqueness or otherworldliness. The ancient Maya are worthy of study and admiration not because they are strange, but because they are altogether human, and because they developed a compelling vision of time unlike that of any other civilization before or since. Their layered but unified philosophy of time and the cosmos no doubt seems very foreign to our way of thinking about the universe, but because of our common humanity, we can approach it, understand much of it, and give credit where credit is due.

I've learned a great deal about the Maya in the course of writing this book, and in fact, in it, I present a few important ideas for the first time. For example, it's become clear to me that published descriptions of the Maya calendar have often been wrong, or at the very least incomplete, even in the academic literature considered the standard reference works on the subject. In chapter 8 I present what I believe to be the first account of how the ancient Maya structured their view of "deep time" using numbers far larger than previously thought. Theirs was a conception of cosmic time that dwarfs anything in our own cosmology by many magnitudes of scale. Many of these new ideas might be half-baked at this early stage, and they will need careful review by colleagues and future students, but I'm excited at the prospects that we may finally be much closer to understanding the big picture of how the Maya and their fellow Mesoamericans "made" time, and created a calendar of remarkable insight and mathematical skill.

Although it might be surprising to some readers, this is *not* a book about ancient Maya astronomy. There are reasons for this. First, despite having a long career in Maya archaeology and epigraphy, I'm far from being an expert in the arcane study of ancient Venus tables, eclipse records, or moon ages—and, frankly, all those numbers make my head hurt. Several colleagues know far more about those things than I do, and have written clear and accessible treatments that I could

never improve upon. But the main reason I've avoided a focus on astronomy touches on a larger point I hope to make: it's simply that the
Maya and their Mesoamerican neighbors, while masters at watching
and understanding the night sky, have for far too long been singled
out in this area as strangely precocious in having advanced knowledge of planetary cycles and the like. Astronomy has been a focus of
Maya research—sometimes *the* focus—for a very, very long time. In
fact, much of the work on Maya art and hieroglyphs back in the 1920s
and '30s involved little else. In playing down the themes of astronomy
in these pages, I in no way wish to deny their importance, but I do hope
to shift the conversation away from the night sky somewhat, at least
momentarily, in order to show that the Maya calendar and the time
concepts underlying it had many other interesting dimensions.

I also want to stress that this book isn't about critiquing the various ideas and theories among New Age thinkers, the so-called "2012-
ologists," concerning that year. These pseudo-scholars purport to know
what will happen, and they even claim some special knowledge about
the Maya, but the published works I have seen are neither very sophisticated nor informed by even a cursory knowledge of Maya archaeology,
history, or culture. I've learned from experience that a point-by-point
refutation of their oddball theories would probably fall on deaf ears, for
scholarship and understanding of Maya culture are not really what's
of interest to them. Instead, their ideas seem to originate from a very
different kind of mentality, one that is often self-centered and rooted
in complex and varied agendas about the challenges and worries of the
present day: the environment, politics, and a quest for inner spirituality. So, for me, a great many of the doomsday predictions and radical
ideas about 2012 say far, far more about the tensions pervading our
own society and culture than anything about the ancient Maya.

This book looks at how scholarship of the twentieth century,
influenced by those ideas of the Maya as so strangely "other," steered
their vision in the popular imagination toward the place we now find
ourselves, where the ancient Maya are somehow seen as strange har-
bingers of our own future. In sum, what follows is really about time
and its place in Maya culture. It's also about how Western academics

and popular culture have struggled to understand it. It examines history, ancient texts, modern Maya religion, and the early development of research to show how the Maya conceived of a remarkable structure to time and space that's significant on its own as a compelling human achievement.

David Stuart
Austin, Texas
September 8, 2009
12.19.16.12.0 13 Ahaw 18 Mol

I

THE ITZÁ PROPHECY

Mac to ah bovat, mac to ah kin bin tohol cantic u than uooh lae?
Who will be the prophet, who will be the priest who shall
interpret truly the word of the book?
    —*Closing statement of The Book of Chilam Balam of Chumayel* [1]

In accordance with the ancient prophecies, the world came to an abrupt end on March 13, 1697. Not our world, of course, but that of the Itzá people and their leader, a king named Kanek'. They were the last holdouts of the ancient pre-Columbian world, until then too remote to be overrun by the Spanish armies and conquerors. But at the end of the seventeenth century, after decades of sporadic war, engagement, and negotiation, Kanek' (a name meaning "Snake Star") finally succumbed to the Spanish forces, which led to the overthrow of the last independent native kingdom in the Americas. For a civilization and a way of life that had lasted millennia before the arrival of Europeans, the defeat of Kanek' was, to borrow a phrase sometimes heard today, truly the "end of times."

The Itzá were a Mayan-speaking people who occupied a large, remote region in what is now northern Guatemala. In the seventeenth century, as for centuries, this wild and forested territory was a refuge for Maya peoples who had escaped Spanish servitude in Yucatán, to the north, and in populated areas of Guatemala, to the south. The forests were dotted with ruined and abandoned cities of earlier Maya civilization, surrounding a large beautiful lake where the Itzá kings lived and held court for centuries. Lord Kanek' and his predecessors had ruled over this territory since the mid-1400s, when, according to native history, the Itzá left Yucatán and abandoned their great capital, Chichen Itzá, today a popular tourist destination. The story of their eventual conquest in 1697 is a remarkable tale, a barely known episode of American history that we are compelled to examine in close detail. For what is most striking about it—in my mind, at least—is that during the decades leading up to their fall, the Itzá Maya had *prophesied* their own end. Numerous written accounts tell us how Kanek' had a strong sense of his inevitable defeat, when, according to the sacred prophecies, a new era of the Maya calendar called a *k'atun* would begin. If we're to believe the firsthand accounts of what happened—and there's no reason not to—native prophecy played as much a role in the defeat of the Itzá as the political machinations of the time, or as did the superior weaponry and numbers of the Spanish soldiers. For the Itzá, the year 1679 was the inevitable "end of an era," a foretold turning point when time would fold and repeat itself, and when political change and transformation were utterly unavoidable. In a fascinating twist on how we usually look on the course of history, native prophecy made the Itzá look upon their conquest and conversion to Christianity as a foregone conclusion.

Early Visits

Incredibly, the final conquest of the Itzá Maya came nearly two centuries after the famous overthrow of the Aztec Empire by Hernán Cortés in 1521. Today, influenced by the well-read and dramatic accounts of Motecuhzoma II's defeat, we tend to think of the "Spanish Conquest" as a single transformative event, a quick and decisive blow, which leaves us with a familiar but simplistic notion of a "pre-Cortés" and a "post-Cortés" Mexico and Mesoamerica. But this is wrong. Whereas defeat and subjugation came quickly amid the remnants of the Aztec Empire—what the Spanish came to call Nueva España, or New Spain—some areas of the Mesoamerican periphery, especially in the Maya region to the east, saw protracted conflicts with native kings and their scattered domains. The conquest of the Maya in Yucatán was in place by 1546, after several poorly planned expeditions met firm resistance from various Maya lords. But far to the south, in the jungles of what is now northern Guatemala, the Maya group known as the Itzá held out, aided by their distance and isolation from areas of Spanish rule. Their capital was an island city called Nohpeten (or sometimes Tayasal), situated on a small island in a vast lake today known as Lake Petén Itzá.* For decades and centuries the Europeans made many attempts at engagement with the Itzá, but difficult terrain made even direct contact daunting and dangerous. So, by the end of the seventeenth century, almost two hundred years after Cortés, and far away from the Aztec capital, the unconquered Itzá still ruled over tens of thousands of subjects, none of them converted to the Christian faith.

As the years passed, settlements of Spanish and Christianized Maya expanded all along the frontier with Itzá territories, leading to numerous violent encounters. During the 1600s, Yucatán and Guatemala were the two main political and economic centers of Spanish authority found south of New Spain, but the Itzá lay directly between them, preventing the construction of an official road—a *camino real*—and the realization of all of the economic benefits that would result. Spanish forces repeatedly attempted to forge this valuable link, often with the agendas and methods of religious and military authorities at odds with

---

* The place name Tayasal has long been used to refer to the island, but perhaps incorrectly. *Tajitzá* means "At (the Place of) the Itzá," in reference to the local ethnic name. In early accounts from the sixteenth century, writers such as Hernán Cortés and Bernal Díaz del Castillo used the corrupted forms *Taiza* or *Tayasal,* and the latter of these remains in customary usage as the old name for the modern town of Flores, Guatemala. Here I prefer to use the more accurate and original Mayan name Nohpeten, "Great Island," based on the preference expressed by Grant Jones, the historian most expert in the story of the Itzá conquest. See Jones, *The Conquest of the Last Maya Kingdom* (Stanford: Stanford University Press, 1998), p. 29.

one another. Would the desired conquest of the Itzá lands be one of brute force, or a peaceful effort of conversion spearheaded by Franciscan missionaries?

Despite being the last holdout of Maya rule, the Itzá kingdom had been visited by Spanish adventurers long before 1697. The very first visit to Nohpeten was by none other than Hernán Cortés, conquistador of the great Aztec Empire. In 1525, fresh on the heels of his defeat of emperor Motecuhzoma II, Cortés and his exhausted army passed through the same dense forests of the Itzá territories, bent on reaching the Caribbean coast of Honduras. There, a rogue Spanish officer had established a rival seat of Spanish power, and Cortés, as de facto leader of New Spain, could not stand for such open resistance to his own rule. His jungle march was one of the most grueling ever recorded in the annals of European conquest, although today its gripping narrative is strangely absent from many history books. Once at Nohpeten, Cortés was suspiciously received by the Itzá king, who was also named Kanek', but the ruler finally allowed the ragtag foreigners to pass eastward through his lands. Cortés's men, weakened but determined to reach the ocean, relied on the Itzás' tentative hospitality for their survival, and had no serious intention to topple Kanek' and his small territory so distant from Mexico City. According to Cortés's own firsthand account, the two men got along well, and Kanek' provided vital information about the lay of the land and the rebellious Spaniards on the coast.*

In preparing to leave Nohpeten, Cortés realized that one of his precious horses was lame, injured by a large splinter in its foot. He was forced to leave the animal with Kanek', who promised Cortés he would take care of the utterly exotic, unknown animal. (Horses were known only in the Old World before the arrival of Columbus in the New.) According to later historical accounts, the Itzá came to worship the

---

* Cortés's brief description of his visit to Nohpeten comes from the set of five letters he composed to the king of Spain justifying his actions, which had already stirred controversy within the Spanish court. The description of Kanek' and the Itzá kingdom are in the Fifth Letter. See Anthony Pagden (trans.), *Hernan Cortés: Letters from Mexico* (New Haven: Yale University Press, 1986), pp. 373-77

lone horse as a divine being, naming it Tziminchaak, and as a deserving god, it was properly fed flowers and birds by the Maya priests.* Predictably, the unfortunate animal soon starved to death. The memory of it lived on once the Maya made a large stone image of the horse and enshrined it at Nohpeten, venerating it for decades to come.

Cortés's description of the Itzá kingdom and its island capital, with gleaming white temples and palaces, planted visions of conquest and conversion in the minds of later Spanish priests and soldiers. Yet decades would pass before Nohpeten would be visited again. The years of protracted fighting and conquest in Yucatán took far longer than in central Mexico, and many Maya simply fled into the frontier zones to the south, toward the Itzá region, where isolation and harsh terrain kept the Spanish largely at bay. Sketchy accounts of the Itzá and their king, Kanek', persisted, and many friars knew that in the remote jungle they would still find thousands of non-Christian souls ripe for conversion, ready "to be friends of the Spanish and to accept the faith of Jesus Christ."

This led to a number of exploratory missions, or entradas (entries), by small parties of intrepid priests bent on converting the remaining un-Christianized souls. The use of the word *entrada* suggests some sense of a crossing over from one place into an utterly different world, which truly must have seemed the case to the friars who, having lived for years among Christianized Maya culture in Yucatán, had no experience of the native, unconquered cultures that still practiced human sacrifice and worshiped various idols in stone temples. At first each of these expeditions of conversion followed well-trodden roads, passing through familiar territory and towns populated with natives long ago

---

* The name of the deified horse, Tziminchaak, offers a telling clue about how the Itzá must have interpreted the strange sight of Cortés and his army. *Tzimin* is Yukatek Mayan for "horse," although its original meaning was "tapir." Chaak is the name of the ancient axe-wielding god of storms, thunder, and lightning. One source tells us that the noise of the Spanish guns was likened to thunderclaps, suggesting that the Itzá may have viewed some of the Spanish horse soldiers as thunderous, Chaak-like deities. Thus Tziminchaak is "Chaak's Tapir," or, more loosely, "Thunder's Horse."

converted to the Catholic faith. But as the days passed, the villages and cornfields grew less and less frequent, the forests became denser and higher, and the roads changed to rustic trails beneath a dark canopy of green. Farther on into the jungle—the priests were seldom sure how far—they would eventually reach the home of the Itzá Maya. But what would they find once there? Memory of Cortés's expedition of decades before was by this time vague and remote in time, and the murky accounts of the island city probably made it seem more like a fantastical legend than a real place.

One of the early entradas into Itzá territory occurred in September of 1618, when Fray Bartolomé de Fuensalida and Fray Juan de Orbita, both Franciscans, traveled for weeks through the jungle, often barefoot, to Nohpeten. Full of zeal and ambition, the two had arrived in Yucatán from Spain only three years earlier, quickly adapting to the ways of the strange New World, including attaining a remarkable fluency in the native Mayan language Yukatek. (The Itzá tongue was nearly identical to Yukatek.) Orbita had already visited Nohpeten in the previous year, apparently making considerable inroads in converting many Itzá, including with their king, Kanek'. Not only did the ruler express openness to adopting the Christian faith, but he had even ordered that a cross be erected beside his royal residence.

Now Orbita brought along Fray Fuensalida as his new companion, and together they ventured into the heart of the "barbarian lands and people." The two were fully expecting to convert more of the king's lords and relatives. Upon their entrance into Nohpeten, surrounded by the king and many of his companions, Orbita was surprised and frustrated to encounter a far colder reception this time to their sermons and lectures, presented in a broken but passable Yukatek. The Itzá lords were unimpressed, emphatic in saying "that it was not time to be Christians (they had their own beliefs as to what it should be) and that they should go back where they had come from; they could come back another time, but right then they did not want to be Christians."

Later that night, in a private audience with Kanek' in the royal residence, the two Spanish priests once again pressed their case about the Itzá accepting Christianity. In their written account, Orbita and

Fuensalida vividly describe how the Itzá king responded: "It is not yet time to abandon our gods. . . . Now is the age of Three Ahaw." Kanek' went on to explain. "The prophecies tell us the time will yet come to abandon our gods, years from now, in the age of Eight Ahaw. We will speak no more of this now. You would best leave us and return another time." At this crucial moment, as they listened to the king's careful words of rejection, the friars' sense of frustration must have been acute.

Their failure to convince Kanek' was in some way predictable, given one episode that had taken place earlier that same day in a temple of Nohpeten. Within one of the shrines near the center of the island, the two friars came upon a vaguely equestrian-looking statue or idol and were told that this was a being named Tziminchaak. Incredibly, it was an image of the horse that had been left behind by Cortés and later deified by the Itzá. Unfortunately, in a fit of zealous rage—or frustration over the Itzás' rejection of his pleas to convert—Orbita destroyed the horse idol on the spot, causing great consternation among the people of Nohpeten, who for nearly a century had remembered and venerated the strange beast as a sacred being.

The tide turned, and soon the priests had no choice but to go. Rumors were circulating that other Itzá nobles who were hostile to Kanek' were amassing warriors and plotting to kill the unwelcome intruders. Hearing the king speak of the power of prophecy among the natives, Fuensalida and Orbita saw the writing on the wall and chose to return to Christian territory. Fuensalida headed for Yucatán, leaving Orbita in a frontier village in what is now Belize. To his superiors Fuensalida took word of his meetings with Kanek', including news of the king's curious and tantalizing prophecy about the strange age called Eight Ahaw, when the Itzá might well accept a new faith.

What Fuensalida and Orbita may not have fully understood was that they had arrived at Nohpeten at the precise moment of a remarkable change in the native Maya calendar, at the turn of a four-hundred-year period known as a *bak'tun*.* Each bak'tun was

---

* As we will explore in great detail, this bak'tun period, ending on September 18, 1618, would be transcribed in the Maya system as 12.0.0.0.0 5 Ajaw 13 Sotz'.

composed in turn of twenty smaller periods, or "ages," called k'atuns, each believed to have its own personality, characteristics, and effect on historical reality. When Kanek' spoke of the time periods called "Three Ahaw" and "Eight Ahaw," he was referring to the names of particular k'atuns, indicating that conversion to Christianity was not in the cards, according to his own priests, but might well be appropriate someday in the future, once the age of Eight Ahaw came to pass. The historical record of the king's conversation with the priests makes no mention of the higher bak'tun period, but it can't be coincidence that within days of the great period's ending, he brought up with his foreign visitors the topic of time, calendar periods, and their influence on the course of human events, at least within his world as king of the Itzá Maya.

We don't understand why Kanek' seemed so receptive to Christianity in 1616 or 1617, yet adamantly rejected it when Fuensalida and Orbita made their entrada in 1618. Why the change of heart? It's possible that he was determined to humor Orbita on his initial trip simply to get rid of an unwelcome intruder. It's also possible, as we will come to see later in this chapter, that "conversion" was little more than superficial in the case of a rather open-ended native religious system that was inherently capable of adopting new elements and fusing or redefining new gods out of old ones. We may never know the true answer. Also, the Itzá kingdom was rife with internal conflict, suggesting that the king was "flip-flopping" his position with the foreigners in order to placate factions within his own court. There's the likelihood, too, that Kanek' had carefully consulted his calendar priests in between Orbita's two visits, learning of a distant future that would see great religious change. Whatever the case, the Spanish priests left Nohpeten and took with them this vital intelligence about the potential power of a native Maya prophecy that the Itzá themselves recognized as valid, foretelling of a conversion of their own kingdom. If nothing else, this gave their fellow clergy an idea of a later return to Nohpeten, when perhaps the Itzá would realize, by their own sacred texts, that the time for conversion was at hand.

The Journey of Fray Avendaño, 1695

Eighty years after Fuensalida and Orbita left Nohpeten in frustration, the Itzá Maya remained independent from Spanish and Christian rule, and another ruler named Kanek' reigned over the island capital. Warriors from Itzá settlements and other independent Maya groups were routinely raiding against the Spanish frontier. Over the years, a handful of European priests ventured forth to Nohpeten, following in the footsteps of Fuensalida and Orbita, but they all met firm resistance from Kanek' and his allies, sometimes with grim violence and death being the result.

In one particularly grisly case, in 1623, a priest named Diego Delgado led a party of Spanish and Maya supporters to the Itzá capital on the lake and were immediately taken prisoner. Friar Delgado was promptly told that he would die in retaliation for Orbita's smashing the images of their gods, including that of Cortés' horse, just a few years before. More than ninety Spanish and their Maya allies were sacrificed in one day, the last being Friar Delgado himself: "they tore open the chest of Father Diego and took out his heart, offering it to their idols in reparation for the offense which they said the other priests had committed. Until that moment he had been preaching to them in a valiant spirit."[2]

As more years passed, pressure to pursue relations built on both sides of the frontier. The Spanish were tiring of the incessant raids, and commercial and political interests led to new efforts to build roads through the remote forest, connecting Yucatán with Guatemala. For the Itzás, trade through intermediaries had increased throughout the seventeenth century, and the various rulers named Kanek' saw the need for machetes and other desirable items from the European world. Intense infighting among Itzá lords pressed Kanek' to form outside alliances and obtain technologies he could use to his advantage. So, in 1695, the king sent his nephew, one Aj Chan, all the way to distant Mérida to make contact and "make a covenant and establish peace between the Spaniards [and the Itzás]."[3] The king seemed eager to establish new and peaceful connections with the Spanish, much to the chagrin of other Itzá nobles, who insisted on continued resistance and war. With such infighting among the Maya, the policies regarding engagement with the Spanish became a true wedge issue in debates, weakening the ruling infrastructure of a kingdom already threatened by Spanish encroachment.

In the early winter of 1695 another party of Spanish priests set out from Mérida, capital of the province of Yucatán, and headed south into the dense jungle wilderness, again toward the Itzá capital. Fray Andrés de Avendaño y Loyola was convinced that things would be different this time. An energetic and determined Franciscan, Avendaño had spent months carefully planning their entrada into unconquered lands. He was a student of native Maya culture and religion, and spoke the Yukatek Mayan language well, if not fluently. Avendaño organized the journey with a single purpose in mind: to convert the natives through the "the destruction of diabolical ammunition which storms the souls of such an infinite number of the heathen as live in those uncultivated wilds." The remote Itzá had long remained an organized and formidable threat to Spanish authority on the frontier, and a potential source of rebellion for neighboring Maya groups who still resisted the rule of the *tzuloob*, the "foreigners." Most galling for Avendaño was that the Itzás still openly worshiped their traditional gods, rejecting the word of Christ. It was too much to take, but now Avendaño had a new and powerful weapon: time itself. After all, the heathen prophecies he had studied over and over again were clear that a new age was about to dawn. He was determined to complete Fuensalida and Orbita's earlier efforts and convince Kanek' that the proper time of abandoning the old gods had finally arrived, once and for all.

Like other priests who lived and worked among the natives of colonial New Spain, Avendaño might be seen today as something of an anthropologist, keen to learn the native language and the rites and rituals he was determined to wipe out. For him, such knowledge of pagan ways was a tool: if he could communicate with the unconverted Maya in their own tongue, knowing their symbols and philosophies, he would be well armed to argue the merits of the Catholic faith. Avendaño had probably practiced his spoken Maya for months in preparation for his journey, but his self-education in native culture went well beyond this. While in Yucatán he had immersed himself in old documents written in the native language, combing through details of traditional Maya lore and the esoterica of the native calendar. In the church archives in Mérida and Izamal, he apparently consulted copies of native documents relating the interlaced stories of ancient history and future prophecy, among them, remarkably, hieroglyphic books that already must have been very old by the late 1600s. His zealous colleagues had not burned all of the native books, and Avendaño probably was secretly relieved to be able to study the old prophecies, and turn them to his advantage.

In the archives, he also may have read and reread Fuensalida and Orbita's report. He pondered their account of their conversation with Kanek' about the power of prophecy, and especially the king's mention that a proper time would come to abandon their gods, with the coming new age called "Eight Ahaw." The two priests' account agreed perfectly with the old native prophecies Avendaño had consulted, which explained that this important new era in the Maya calendar was now fast approaching. In less than two years, in 1697, the Itzá would see the turn of a k'atun, the twenty-year period that formed the backbone of the native system of time reckoning. Its name, Avendaño knew, would be Eight Ahaw.*

Avendaño's party took weeks in its southward journey, passing first along well-trodden roads through Maya villages, then into the rain forest, where people were few in number and trails barely discernible. At one remote outpost named Chuntuki, they were excited to find a narrow trail heading deeper into the woods, more or less in the direction of Itzá territory. According to the priest's account, they took off running down the path, sure that their goal was not far away, all the

---

* Debates are ongoing over the precise time of the turn of the k'atun cycle. Some propose that it had already occurred in 1695, whereas others opt for the summer of 1697, a few months after Avendaño's visit. See Jones, *Maya Resistance to Spanish Rule*, pp. 325-26, for a good discussion of the different proposals offered by Bricker, 1981, and Edmonson, 1986.

while chanting "*in exitu Israel de Egiptu,* in order to imitate the victory of the Israelites, who succeeded in making their way across the waves of the Red Sea." After days of walking into unknown terrain, crossing numerous hills and fording streams, they arrived at a large water hole, or *aguada,* where they came upon the ruined remains of tall Maya temples, long abandoned and covered by trees, but evidently still the sites of pagan veneration, as indicated by the clay idols Avendaño says he destroyed there. The identity of this impressive ruin—one of hundreds in the area today—remains a mystery. Finally they came to a native settlement near the great lake in Itzá lands, the first sign of habitation after nearly a week in the forest. As they approached the center of the village, Maya women and children ran from them screaming, warning the others of the strangers' arrival, and several warriors came out wielding bows and arrows. Avendaño walked up to the men and embraced them, showing off "some things from Castille" that they had brought as offerings and trade goods for the Itzás. Most of these things were carted away by the locals that day, much to Avandaño's dismay. But the peaceful entreaties had their effect, and the travelers were able to spend the night in relative tranquillity, watching dances and ceremonies being performed by the Maya.

The next morning the priests and their entourage made for the shore of the vast Lake Petén Itzá. They arrived and forwarded a message to the king, then waited much of the day for his response. Finally, at two o'clock, the king, named Kanek' like his forebears, arrived on the lake shore in a grand flotilla of eight canoes "full of Indians, painted and dressed for war." Avendaño had expected the standard exchange of formalities—something he had no doubt practiced for weeks—but the king's guards instead whisked him and his men away to the dugout boats and took them across to the island of Nohpeten. The friar understood little of what was happening, but he was more annoyed than frightened, as he tells it, irked by such hasty and undignified treatment. The Itzá king had neither regard nor appreciation even for the "music of the clarions with which we awaited him."

A remarkable exchange then took place in the royal canoe, as Avendaño and Kanek' crossed the lake together, heading toward the island city. Avendaño's own words provide the best account of what happened next:

> Suddenly the king placed his hand on my heart to see if it was at all agitated, and at the same time he asked me if I was so. I, who was before very glad to observe that my wishes and the work of my journey were being realized, replied to him, "Why should my heart be disturbed? Rather it is very contented, seeing that I am the fortunate man who is fulfilling your own prophecies, by which you are to become Christians; and this benefit will come to you by means of some bearded men from the East, who by the signs of their prophets we ourselves, because we came many leagues from the direction of the east, plowing the seas with no other purpose than to bring them, borne by the love of their souls (and at the cost of much work), to that favor which the true God shows them."
>
> I at this time, with some liberty on my part, also placed my hand on his breast and heart, asking him also if his was disturbed, and he said: "No." To which I replied: "If you are not disturbed at seeing me, who am the minister of the true God, different in everything from you—in dress, customs, and color, so that I inspire fear in the devil, and if your heart is not troubled, why should you expect me to be afraid of you, mere men like myself, whom I come purposely to seek, with great pleasure, merely for the love which I have for their souls, and having found them, in order to announce to them the law of the true god, as you shall hear when we come to the Peten."[4]

Soon the royal canoe landed on the shore of a steep circular island covered with temples and houses. Avendaño was escorted throughout the town, and the curious Itzá people soon emerged in droves, "coming up to the high places so as to see us; all proofs of joy and wonder."

At last the European priests, still surrounded by scores of curious gawkers, were led up the hill to the king's residence. Before its entrance stood a masonry table, a sacrificial altar, which immediately filled the priests' hearts with great trepidation—would they be put to death in honor of the destroyed idols, or otherwise murdered, like some earlier missionaries? Such fears were put to rest when the friars found themselves being given food and drink in a darkened chamber, which, they were relieved to discover, was the "welcome room" of the king's palace. It was also, on this special occasion, an interrogation room, where Avendaño could barely see his inquisitors. "Why did you come to Tajtzaj?" the king and his officers asked. Ever prepared, Avendaño reached for the formal letter of introduction that would explain all, as well as a message he had carried from the governor of Yucatán. Unable to see or read anything in the dark chamber, he requested to go outside to proclaim the purpose of his visit. The king granted the request, and together they exited into the bright sunlight.

Kanek' escorted them to the largest, highest temple on the island. Inside, the European priests took particular notice of a curious box that hung suspended from the shrine's ceiling, "in which we saw . . . a bone leg or thigh, very large in size, which appeared to be that of a horse." In shock, Avendaño and his companions suddenly realized that this was the bone from the very animal left by Cortés, "which they had kept as a relic or to hold him in memory." Fray Orbita had destroyed the idol of Tziminchaak decades before, yet, incredibly, the true remain of Cortés's horse was still venerated nearly two hundred years after the great conqueror had left the animal in the care of the Maya.

Avendaño then "brought out the letters . . . and it cost no little trouble to make them sit down and keep quiet." Before the kings and his own priests and officials, Avendaño read his papers, each translated word for word into Mayan from the original florid and formal Spanish of the time. They were lengthy statements of intent, explaining how the priests had come from far away to reveal the word of God in accordance with the Itzás' own ancient prophecies. But after the reading there was little response from his audience, only many confused expressions. Avendaño asked his audience what was wrong. "We do not understand what you say," responded the Itzá. It was clear that the letters, while using Mayan words, were simply too odd-sounding, couched in the alien tone and structure of a formal European discourse. In frustration, the friar put the letters aside and chose to speak ad-lib, "explaining the said message to them in the ancient idiom."

His new strategy worked far better. The native king and his people finally understood, apparently, now that the words were stripped of their European phrasing and style. Finally, as the king and his gathered lords rose to leave the temple, Avendaño pressed for a substantive response to his mission: Would they convert and cleanse their souls? The Maya answered by saying, "Not yet. We will think about it first, for there is time in answering. Wait."

If we take Avendaño's account at face value, the Maya were particularly surprised and impressed by his mention of the ancient prophecies and by his ability somehow to interpret them. It all left a deep impression on Kanek' and his priests. In their eyes, surely, prophecy and the meaning of their calendar were highly esoteric and guarded pieces of knowledge, kept closely by the native temple priests. To hear a Spanish priest recount them in the native language must have been an especially jarring experience. The Itzá, Avendaño wrote, "began to love and fear me at the same time, saying I was undoubtedly a great personage in the service of my Gods, since I had succeeded in learning the language of their ancestors." The Maya even gave Avendaño two new special titles of honor, dubbing him "great lord" and a "father of heaven." They could not leave the strange foreigners alone—throngs came to see the Spaniards all day and all night, leaving the friars no privacy. "Neither the prohibition of the King nor our own scolding was sufficient to hinder their excessive curiosity, the only attention they paid to either being a lot of laughter."

Over the next days it was clear that Avendaño's telling of the prophecies had hit home with the king. Obviously he had been aware of them, as had his predecessors on the throne, so the priest's points were really nothing new to Kanek'. But he nevertheless seemed to gain assurance with Avendaño's adamant words. Within days, the Maya king and two of his priests agreed to have themselves baptized. Still, the Maya pressed Avendaño about the strange baptism ceremony and what it would entail. When deep in discussion on the issue, the king offered his young son to the priest, so that all could see a demonstration of just what baptism looked like. Satisfied upon seeing the application of holy water upon the boy, Kanek' and several of his entourage agreed to take part in the ritual, receiving the word of God and new baptismal names.*

Things soon went wrong. During the visit of Avendaño's party, several prominent Itzá lords and the king's own wife had been plotting to expel the priests and put the brakes on the king's defeatist and fatalistic attitude, now clear for all to see. They saw the king's close relation to the Spanish as troubling, to put it mildly, and his interpretation of the prophecies as nothing less than disastrous. Tensions among the Maya factions ran high until, in a replay of Fuensalida and Orbita's ungracious exit decades earlier, Avendaño and his fellow priests were forced to leave. The Spaniards slipped out under the cover of night and made their way toward the Spanish frontier settlements in what is now northern Belize.

By this time, though, the stage had been set for the end of the Itzá. Patience among the political authorities in Yucatán had now run out, and in March of 1697, much to the chagrin of the Franciscan friars, Spanish troops finally arrived on the shore of Lake Petén Itzá from Yucatán. Whereas Avendaño had presented Kanek' with peaceful terms for conversion and alliance, the secular governor of Yucatán, Martín de Ursúa, was determined to move forward on the conquest of the Itzá region. He, too, was apparently aware of the Maya prophecy and seemed bent on seeing it through to its inevitable conclusion. Hundreds

---

* In the complex history of the conquest of the Itzá, we find that the conversion of Kanek' was mirrored by an earlier ceremony that had taken place in far-off Mérida, where Avendaño and his companions began their journey. Not long after their departure, a group of Itzá emissaries visited Mérida in order to gauge the possibility of forging peaceful ties with the Spanish to the north. For years the Itzá had felt beleaguered by military engagements originating from Guatemala, to the south, and felt that Yucatán offered a far more advantageous option, given the apparent inevitability of closer associations with the Spanish. On December 31, 1695, the king's ambassadors, including his nephew Aj Chan, agreed to be baptized, all probably aware of the nature of their own prophecies. See Jones, 1998, for a complete discussion and compelling analysis of the complex power politics that underlay the religious conversions of the Itzá.

of troops set off from the lakeshore for Nohpeten, some in a large boat bearing a cannon. But the great battle of conquest never came; when they arrived, they found Nohpeten almost empty, its houses, temples, and palaces largely abandoned. The soldiers descended on the shrines and began destroying various idols and pagan symbols. In one temple, they came across pieces of old horse bones. A frail old Itzá woman left behind on the island told them these were the bones of Tziminchaak, the horse given to them by the great conqueror Cortés nearly two centuries before. Soon the temples were razed, and churches were hastily built in their place. Nohpeten was renamed the island of Nuestra Señora de los Remedios y San Pablo, sealing the fate of the Itzá kingdom. (Flores would become a later designation for the island, one used to this day.) Not far from the lake, Kanek' was taken prisoner by Spanish soldiers. We know little of what happened to him in later life. The final mention of the once-great leader informs us that after his capture he was taken to the capital of Guatemala, the colonial city today known as Antigua, where he lived out the remainder of his days.

Conversion and Prophecy

The quick conversion of Kanek' may seem too pat, too easy. Was it heartfelt and genuine? It's very hard to know, given the very important political and economic motivations behind his effort to make alliance with the Spanish crown. Nevertheless, there's no question that native prophecies were a key element in his decision-making, if not a determining factor. For the Itzá and other Maya, time was not just a means of measuring the course of history but was also a deterministic, shaping force in human experience. Whereas we are mostly accustomed to seeing time as utterly impersonal and abstract, Maya and broader Mesoamerican notions of time show it was very much an idea, an "actor" invested with personality and character who shaped the very nature of history itself. It was the passing of the k'atun—one of many artificial and man-made constructs of categorizing time—that finally convinced the later Kanek' of the futility of continuation. In the minds of Maya kings, time's character evidently held more power and influence than they themselves did. The role of the ruler was to harness and manage time's unremitting movement, to live and rule according to its mechanisms and its overarching authority. Like his ancestors, the younger Kanek', who ruled at the cusp of k'atun eight Ahaw, therefore was destined to follow time's larger meaningful pattern. He "chose" to convert himself and his people to the Christian faith, even if fate had much to do with his decision.*

But was it a true rejection of one old religion in favor of a new one? The Mesoamerican understanding of religious conversion differed considerably from what Catholic authorities had in mind. The early years of contact between natives and Spanish show that Indians would often easily embrace Christian symbolism and ideology, desirous to add them to their repertoire of sacred images to be venerated. A wonderful illustration of this was recounted in the famous narrative of Bernal Díaz del Castillo, one of Cortés's soldiers, who witnessed one telling example of a supposed "conversion" among the Indians of Tabasco while on the way to conquer the Aztec capital:

> One other thing that Cortés asked of the chiefs and that was to give up their idols and sacrifices, and this they said they would do, and, through [the interpreter] Aguilar, Cortés told them, as well as he was able about matters concerning our holy faith, how we were Christians and worshipped one true and only God, and he showed them an image of Our Lady with her precious Son in her arms and explained to them that we paid the greatest reverence to it as it was the image of the Mother of our Lord God who was in heaven. The Caciques replied that they liked the look of the great Teleciguata (for in their language great ladies are called Teleciguatas) and begged that she might be given them to keep

---

* There were probably also important political reasons behind the king's decision. Jones, 1998, also offers an expert analysis of the complex factions and politics that existed in the final years of the Itzá kingdom, and how these played a part in the ruler's difficult choices.

in their town, and Cortés said that the image should be given to them, and ordered them to make a well-constructed altar, and this they did at once.*

So much for the idea of native resistance—at least for now. While the first sight of Christian pomp and ceremony must have seemed strange to the Mesoamericans hosting Cortés, it seems that aspects of the visitors' ceremonies had some familiar ring to them. The costumed priests, the performance of ritual with chanting and the burning of incense, all focused on an impressive sacred image, clearly resonated with traditions long familiar to the natives. While certainly a biased source in many respects, Díaz del Castillo's description of the enthusiastic adoption of the Virgin Mary's image rings true, I think, for it reflected a fundamental "nonexclusive" concept in Mesoamerican spirituality and religion. In a world where there were multitudes of gods and sacred images, adding another exotic one to the mix may not have been too big a deal. The Spanish, of course, had no such flexibility in their notions of religious change. For them, the thousands of baptisms in the first decades of New Spain represented the ultimate spiritual victory, an all-or-nothing proposition. And yet they were very much unaware that the new religion they brought to the New World was being merged with long-held native religious traditions, rather than replacing them outright.

It's likely, then, that Kanek' and his Itzá allies saw the Spaniards' proposals, and perhaps even their own prophecies, not as a signal for a complete switch from one system of belief to another. "Conversion" was perhaps never the idea. Instead, based on what we can discern about Maya notions of prophecy, the Itzá probably foresaw the turn of the cycle more as a *recurrence* of some previous historical era or experience, familiar and predestined according to the patterns of native history.

---

* Bernal Díaz del Castillo, 2003, p. 63. Bernal Díaz's highly readable and detailed firsthand account of the conquest of Mexico is one of the great classics of historical literature.

The prophecy of the k'atun called 8 Ahaw was shaped by the events and characteristics of earlier k'atuns, from centuries in the past, bearing the same name.

In the calendar used at the time, each twenty-year k'atun assumed its own "personality" and character. There were 13 k'atuns that made up a larger cycle of 256 years, after which the named k'atuns would repeat. Each k'atun period was named after a day in the calendar on which it ended, so that the period called "Eight Ahaw" ended on a day of the same name. (Scholars prefer to transcribe this nowadays as "8 Ahaw.") I will explain the mechanics of that system a bit later, but suffice it for now to know that the k'atun periods used by the Itzá ran concurrently with the Christian calendar, and that they could be correlated in this way:

k'atun	5 Ahaw	AD 1599–1618
k'atun	3 Ahaw	1618–1638
k'atun	1 Ahaw	1638–1658
k'atun	12 Ahaw	1658–1677
k'atun	10 Ahaw	1677–1697
k'atun	8 Ahaw	1697–1717

Each named k'atun would repeat in cycles of thirteen, so that, for example, a k'atun "5 Ahaw" or "8 Ahaw" would occur about every 256 years. Each of these twenty-year "ages" in the past and in the future had their own identity and historical character, or, as Avendaño put it, "each age has its own particular idol and its own priest, with a particular prophecy of events."[5] This cyclical system of time gave rise to the idea that history was forever based on familiar recurring patterns, and that "prophecy," at least in the Maya understanding of time, was but a reflection of events and trends of the past. The Itzá and the earlier Maya of Yucatán revealingly referred to these repeating patterns as "folds" of k'atuns, a term that suggests repetition and overlay. Prophecy, therefore, was a reflection of the past. This decidedly cyclical structure of history gives us only part of the picture, however, and as we will explore in far more detail, Maya concepts underlying the calendar and ideas about the course of history elegantly integrated both linear and cyclical notions of time's progression.

We know a considerable amount about Maya prophecies from Yucatán, since a number of the old historical texts from there still survive for us to study. As Orbita, Fuensalida, and Avendaño knew, Yucatán had been full of native documents relating both native history and prophecy—the two concepts were indeed hard to distinguish in native thought. Numerous such books had been destroyed in earlier efforts to wipe out pagan idolatry, but others had been kept hidden by Maya priests and shamans, often carefully transcribed from hieroglyphic books into the Latin alphabet. Many native communities in Yucatán and elsewhere retained such sacred writing throughout the centuries, and, incredibly, a handful of them have been preserved to this very day. The most important of these documents is a series of manuscripts known as the Books of Chilam Balam, named after a famous native Maya prophet and containing passages describing the historical episodes of past k'atuns. Previous eras of k'atun 8 Ahaw—the time Kanek' specified as a moment of change for him and his people—shared a key historical theme whereby the Itzá left or abandoned their communities. The Books of Chilam Balam contain passages that describe great change and destruction:

> (Katun) 8 Ahau was when Chichen Itzá was abandoned. There were thirteen folds of Katuns when they established their houses at Chakanputun.[6]

In 13 "folds" of k'atuns, about 256 years, we arrive at a similar episode of history:

> (Katun) 8 Ahau was when Chakanputun was abandoned by the Itzá men. Then they came to seek homes again. For thirteen folds of katuns had they dwelt in their houses at Chakanputun. This was always the katun when the Itzá went beneath the trees, beneath the bushes, beneath the vines, to their misfortune.[7]

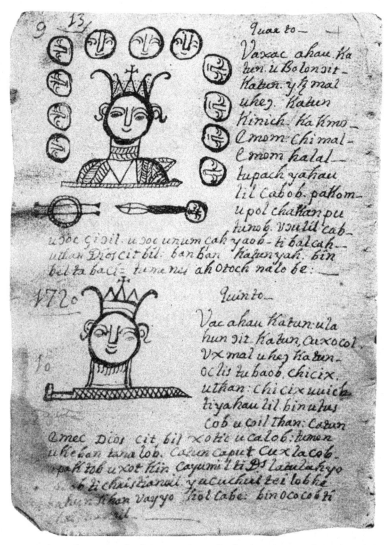

A page from the Book of Chilam Balam of Chumayel, *an eighteenth-century Maya manuscript recounting native history and prophecy. The crowned heads each represent a Europeanized image of a lord* (ahaw) *ruling over a k'atun time period, roughly twenty years.*

And thirteen k'atuns after that, describing some of the murky political intrigue of that era, the Books of Chilam Balam state:

> (Katun) 8 Ahau was when the Itzá men again abandoned their homes because of the treachery of Hunac Ceel, because of the banquet with the people of Izamal, For thirteen folds of Katuns they had dwelt there, when they were driven out by Hunac Ceel because of the giving of the questionnaire of the Itzá.[8]

In the Books of Chilam Balam such statements about past events are clearly juxtaposed with future predictive history, laying out perceived patterns in the characters of the k'atuns. The priest did not simply lay out a series of random predictions about the happenings of the world to come. Rather, prophecy was the outcome of recognizing and interpreting perceived patterns in past historical cycles and extrapolating from them certain projections into the future. There was a careful analytical system to these sorts of predictions, and not so much a passive reception of visions about prophetic history.

Chilam Balam was no pre-Columbian Nostradamus, but who was he exactly? Chilam Balam, or "Jaguar, the Prophet," was the name of a celebrated priest who lived in the town of Mani, Yucatán, near the end of the fifteenth century and the beginning of the sixteenth century.* Tradition held that he alone had foretold the coming of strangers from the east who would bring with them a new religion for the Maya people. For this he was held in great esteem, perhaps as a Maya version of Nostradamus. So great was his fame, in fact, that the native Maya books of prophecy that were compiled in various towns during the seventeenth and eighteenth centuries bore his mark, and came to be known collectively among Mayanist scholars as the Books of Chilam Balam. These loose aggregations of history, prophecy, astronomy, timekeeping, and medical knowledge were not written by the original Chilam Balam, but they nonetheless came to be associated with this

---

* Balam, or "Jaguar," was likely a surname, as it still is throughout Yucatán to this day.

great soothsayer, a symbol of native cultural knowledge and intellectualism, even after two centuries of Spanish rule in Yucatán.

The curious title we translate as "prophet," variously given as *chilan* or *chilam*, was reserved for a class of highly regarded priests and soothsayers who lived throughout many towns in ancient Yucatán.* Far from being remote intellectuals who were closed away in temples, the *chilans* had important functions within their communities, as interpreters and fortune tellers for the people at large. As Bishop Landa describes, "The role of the *chilanes* was to give the replies of the demons to the townspeople, who so highly regarded the priests that they carried them upon their shoulders."† One of the several surviving Books of Chilam Balam, from the town of Tizimin, states that the *chilans* would receive their prophetic words within their own houses, where the priests would retire and lie prostrate on the floor. A god or spirit, perched on a roof beam, would then "speak" to the entranced *chilan*.[9] One wonders if the "perched" spirits in these cases were singing birds, since the word for both "bird" and "omen" is the same (*muut*) in Yukatek Mayan.

So it's unlikely that our own sense of "prophecy"—basically the ability to predict and foretell future events—matches exactly what the Maya had in mind with their notion of k'atun periods folding upon one another, and with each era resonating with others in both the past and future. Time was structured. Events and trends of history repeated themselves from one era to another, and patterns of repeated rises and falls were an essential feature of Maya history as far back as any records show. To the Maya, who composed such books and manuals about the workings of the calendar, it was important to understand that the flow of time and history was *readable*, almost an exercise in pattern recognition.

Stepping away from the Books of Chilam Balam and the Spanish

---

* The original term *chilan* derives from the verb *chil*, "to lie down," in reference to the special position the priests assumed when receiving their prophecies. Some earlier scholars (e.g., Roys, 1967, p. 3) erroneously translate the word as "mouthpiece," seemingly based on *chi'*, meaning "mouth, word."

† My translation here is slightly modified from that by Tozzer, who neutrally translates Landa's *demonios* as "gods." See Landa, 1941, p. 112.

priests' accounts of their speeches at Nohpeten, it's not easy to discern how important a role prophecy played in more ancient Maya religion, history, and philosophy. No hieroglyphic inscription from the so-called Classic period, when most of the famous Maya ruins were occupied, has anything to say about the future, except perhaps to anchor contemporary history to the large-scale cycles of the calendar. Most ancient writings are commemorative texts on stone, relating the ritual actions of kings and their relatives, dedicating monuments and buildings, recounting military victories, and so forth. In flavor and subject matter they are very unlike the k'atun histories of the colonial period, and prophecy simply isn't featured in them. Nevertheless, I suspect that if we could imagine browsing through an ancient library at Palenque or Tikal in AD 700, we probably would come across many hieroglyphic books with language and content not too different from the Books of Chilam Balam. Like the screen-fold pages on which they were written, such historical texts would cleverly interweave notions of time, folding the past, present, and future onto one another, forming a single patterned history. These perishable genres of history writing and recordkeeping simply didn't survive the famous collapse of the Classic Maya around AD 800.

Vague hints of prophecy do exist here and there in the ancient stone inscriptions from the Classic period, however. In one statement that survives in an inscribed tablet in a temple at Palenque, we read of the crowning of a god as a ruler several thousand years in the distant future from *today*. And just as with the "folds" of time represented in the Books of Chilam Balam, this future event is a reflection of the past, linked in the inscription directly to the crowning of Palenque's great historical leader K'inich Janab Pakal, who ruled for most of the seventh century. For the ancient Maya, too, "prophecy" was all about context, a tool for understanding and interpreting contemporary history and politics.

It's in this same light that I should introduce one other important future happening, cited in another ancient Maya text from a small ruin called Tortuguero, not far from Palenque. On a large inscribed slab that once adorned a temple shrine there, we read that "thirteen bak'tuns will end" on the day that, in our calendar, corresponds to

December 21 or 23, AD 2012. This is the single clear mention of the notorious 2012 that some take to be the "end of the Maya calendar."

It's for this reason, nearly four centuries after the fall of Nohpeten, that Maya prophecy is back in the news. Countless books, articles, television shows, and films tell us that the ancient Maya predicted that another "end of times" would occur in 2012, now just a few short years away as I write these words. Many, including me, don't believe a word of it, but given the deluge of press it's getting these days, others are bound to wonder what the ancient Maya could have known about our own troubled times, and whether something dire might be around the corner. As someone who has spent most of his life fascinated by Maya and Mesoamerican culture, spending years deciphering and analyzing their intricate historical and religious texts, I have to scratch my head in wonder. What predictions are people talking about? What prophecies? As later chapters will explain in detail, the Maya calendar doesn't in fact end in 2012, nor did the ancient Maya soothsayers leave us any prophecies of fast-approaching (or even slow-approaching) doom and gloom. The Tortuguero inscription tells us little about what the Maya had to say about the upcoming date, and any claims to the contrary are simply wrong. I will explore that inscription in far more detail in the final chapter, but suffice it to say for now that the Maya calendar has little if anything to do with our modern world—with the important exception, perhaps, of the Maya people who still hold on to elements of ancient timekeeping, and continue to make them an important part of their daily lives.

Having said all of this, I do believe that the story of the fall of the Itzá shows us how it would be a mistake to summarily discount Maya prophetic history as little more than smoke and mirrors, with no real effect on history's course. Prophecy has the power to "work" when enough people or certain influential individuals believe in it, as seems to have been the case of Kanek'. Time doesn't directly affect history, but people's perception of time and its structure certainly can, and did. As we will explore in great detail throughout this book, the native Maya calendar in this way played a key role not only in influencing how the Maya perceived events of the past and future, but also in actively shaping the course of history itself, not just in the instance of the Itzás' self-fulfilling prediction of their own end.

◇

In no sense did the defeat of the Itzá represent the end of the Maya. Speakers of Mayan languages, millions in number, today live throughout the highlands and lowlands of Guatemala, Mexico, and Belize—and the same languages can even be heard in the streets of Los Angeles and in the tobacco fields of North Carolina. Many aspects of traditional Maya religion and culture survive as well, adapting and modifying themselves to the pressures and realities of the modern world. One of the great inexplicable myths in our popular perception of the Maya, in my mind, is that they somehow disappeared or mysteriously left at some point in ancient history. Obviously they didn't, but as with many other Native American peoples, efforts at cultural and linguistic survival haven't always been easy or successful. Many of the thirty or so Mayan languages remain vibrant and healthy, while others face extinction just within the next few years. Tragically, one of these doomed languages happens to be Itzá (or Itzáj, as it's sometimes spelled). Today it's spoken by only a few dozen elderly men and women in the town of San José, on the lakeshore not far from the island that was once ancient Nohpeten. In a few years, inevitably, the Itzá will be truly gone.

Today I regularly visit the island, now called Flores, whenever I go to Guatemala on my way to conduct field research at nearby Maya ruins. I usually walk its charming curved streets and alleys, sometimes trying to imagine scenes of Cortés, Avendaño, and the various kings named Kanek' locked in their intense discussions of politics, religion, and philosophy. Amid the hotels, tourist shops, and cafés, with the bustle of tourists and motorbikes, it is not easy to transport oneself back to that earlier era, really not so long ago. Yet something about the town still conveys, for me at least, a remote sense of sadness and loss, of a native world destroyed yet still somehow present under the cobblestone streets. From time to time, workers digging the streets to lay cables or

The island town of Flores, Guatemala, the site of the ancient Itzá capital Nohpeten.
(Photograph by R. Lopes Bruni)

repair sewers come across smashed fragments of ceramic idols, or even stone sculptures from ancient temples long dismantled. Many pieces of Nohpeten's temples and its palace still remain beneath the alleys of Flores, and I like to think that perhaps someday an archaeologist will come across a few scattered horse bones, or fragments of a Maya carving vaguely resembling a Spanish steed.

While in Flores, I often like to walk the single long street that runs along the outskirts of the island, drawing a perfect circle back to my small hotel by pivoting around the church up on the island's high central point. The cobblestone street of today likely follows the course of an avenue of ancient Nohpeten once walked upon by Cortés, Orbita, Fuensalida, and Avendaño. As I amble around the island, I always think how interesting it is that Nohpeten's round shape and its outer path conform to a very Maya model of the universe, as well as of time and space. We see hints of this in the word *peten,* "island," also meaning "province." This derives from the more basic term *pet,* meaning "round, circular." Maps drawn by native cartographers from the colonial period regularly assume a circular form, as do the ancient calendrical charts we find preserved in handwritten chronicles of that time—what we call "k'atun wheels"—which served as representations of a unified conception of Maya time and directional space. How fitting, then, that the capital Nohpeten (Great *Peten*) was so perfect a circle, a microcosm of space, an experiential wheel of time. There is no proof whatsoever of this, but I enjoy thinking that the circular route that defines the island's edge may once have been important in formal religious processions that replicated the movement of time itself, and in other ceremonies led by Kanek' and his priests.* No one now walking the streets of Flores today ever considers any of this spatial and temporal symbolism, but I have no doubt they were real and important sources of the town's ancient layout. For Kanek' and his people, all now largely forgotten to history, the tiny island of Nohpeten was the center of the universe.

---

* One early description of Nohpeten, written only a few years after the Conquest, emphasizes the religious importance of the island capital: "The Indians, its native inhabitants, called it Noh Peten, which means large island, not because of its material grandeur (as it is small), but rather because on it its ruler always lived and on it they also had the principal temples of their idols and carried out the most solemn functions of their idolatry." This was written by Friar Diego de Rivas, and is quoted by Jones, 1998, pp. 68–69.

2

MESOAMERICAN TIMES

When we saw so many cities and villages ... we were amazed and said it was like the enchantments they tell of in the legend of Amadis. .... And some of the soldiers even asked whether the things we saw there were not a dream.

— *Bernal Díaz del Castillo, soldier in the army of Hernán Cortés, describing their approach to the Aztec capital Tenochtitlan in 1519*[1]

The Itzá realm centered at Nohpeten was the last in a very, very long line of native kingdoms—some large, some small—that had come and gone in the region over the course of the previous three thousand years. The defeat of Kanek' represented the end of native kingship as a long-lived institution and, one can argue, the true demise of ancient Mesoamerican civilization as a whole. For that simple reason, I don't think it's hyperbole to see the Itzá conquest as a momentous and transformative episode in human history. In the tens of centuries before the arrival of Cortés, the vast area that is today Guatemala and Mexico had been home to countless different kings and courts, all part of a constellation of cultures and traditions that we define as Mesoamerican civilization. Some of the names for the peoples ruled by ancient kings are familiar, others less so: Aztec, Mixtec, Zapotec, Mixe, Zoque, Olmec, and Maya, among others. Yet on that one spring day in 1697, that world was finally and permanently over.*

---

* It's important to stress that the Itzá conquest wasn't the end of Mayan independence in the long term. Decades of resistance and rebellion continued in

The remote origins of Mesoamerican civilization are far less easy to define in time and prehistory. As we look deeper into the past, we see that early Mesoamericans followed a trajectory of development and cultural evolution not too different from that in other areas over the world, where demographics, environment, agricultural technology, politics, and religion all interacted to create something we understand as "civilization"—hardly a scientific or objective term, but useful for our purposes here.

Beginning some five or six thousand years ago, several different regions of the world saw independent beginnings of urban life, all characterized by the advent of large towns or cities and the associated complex social, political, and economic institutions. The driving force behind this great demographic change in human history was the cultivation of food and the ability to adapt technology to the inevitable increases in population that came as a result. Agriculture didn't always come first in this equation; in some areas, small human populations could concentrate and thrive simply on the basis of gathering food. But the cultivation of various staple foodstuffs was generally a key in the creation and sustainability of settled life. This in turn led to continual increases in human population in certain regions of the globe. Over centuries and millennia, people grew in number and condensed into ever-larger settlements, necessitating an expansion of economic networks and the diversification of a skilled workforce. By 4000 to 3000 BC, large and sophisticated communities existed in many areas of the world, some having developed earlier than others; all were driven by a universal human tendency to enjoy the varied benefits of a social network.

Mesopotamia—more or less where modern Iraq appears on the map—has long been seen as the "cradle" of all human civilization, with

---

Yucatán, and in Chiapas, well into the twentieth century. One independent Mayan polity in southeast Yucatán lasted through the late nineteenth century before being overthrown by Mexican troops in 1910. The Itzá conquest did, however, represent the end of a pre-Columbian mode of kingship not present in these later political movements.

its very early urban monuments and first indications of writing. But, tantalizingly, there is increasingly clear and vivid evidence that human urbanism arose in equally complex fashion within mere centuries in several areas of the world, including in Egypt, South Asia, China, and the Americas. The archaeological world was stunned not too long ago when buildings at the Peruvian city of Caral (far to the south of Mesoamerica) were dated to about 2600 BC, nearly contemporary with the Egyptian pyramids of Giza. More recent architectural finds at other sites on the Pacific Coast of Peru, such as Sechin Bajo, may date back even earlier.* As archaeological research continues apace, dates for the "first this" and the "first that" are continually being pushed further and further back in time. This suggests that, to some degree, "civilization" had several different beginnings in human history, all within a remarkable window of a few thousand years.

Today we use the word *Mesoamerica* to refer to a series of interconnected cultures that arose in what is now southern Mexico and northern Central America, beginning some four thousand years ago. The term is a relatively new coinage, first proposed by the German Mexican anthropologist Paul Kirchhoff in the middle of the last century.† By 1950 or so, scientific research on the Maya and their neighbors had accumulated enough good information to make clear that many of the cultures of the region shared a number of key features not found elsewhere. Kirchhoff's original laundry list of cultural features included an emphasis on maize agriculture; the widespread use of two related calendars, one of 260 days and another of 365 days (to be described in detail in chapter 5); a religious tradition that involved strong concepts of sacrifice; the use and construction of stone pyramids; the playing of a distinctive ball game,

---

* For the early Peruvian site of Caral, see Shady et al., 2001. The work on Sechin Bajo has yet to be published in full, but it is described in some recent press releases. Online at http://news.nationalgeographic.com/news/2008/02/080226-peru-oldest_2.html

† See Kirchhoff, 1943. Kirchhoff studied and taught in his native Germany for many years, but his deep involvement in leftist politics during Hitler's ascendancy led to his departure for France and then Mexico in the mid-1930s. He taught at the National Autonomous University of Mexico (UNAM) until his death in 1972.

General Map of Mesoamerica showing major ancient sites and cultures.
(Drawn by Philip Winton for *The Art of Mesoamerica: From Olmec to Aztec*
by Mary Ellen Miller, Thames & Hudson, London and New York)

the courts of which are found adjacent to large ceremonial structures; and the widespread use of "pictographic" writing, found on various media, including bark paper manuscripts, or on large stone monuments.

By weaving together various strands of knowledge, Kirchhoff validated and codified what many scholars of the era already knew full well: that the Aztecs, Maya, Zapotecs, and other neighboring peoples were essentially variations on a similar cultural theme, all related to and perhaps derived from a common ancestral culture of great antiquity. He opted to call these connected cultures "Mesoamerica," and the name has stuck ever since.

Interestingly, some of the traditional features of Mesoamerican civilization appear well outside of its geographical confines, even among ancient North American cultures. For instance, evidence of the same ritual ball game has been found at sites in the American Southwest, including the ancient cities of the Hohokam culture, located in present-day Arizona; they must certainly have traded with their neighbors to the south. Most remarkable, too, are the large earthen pyramids and plazas built between 200 BC and AD 1500, in elaborate ceremonial centers such as

Cahokia, near St. Louis, Missouri, and Moundville, located in the state of Alabama. All of these cultures to the north prospered with the cultivation of maize, and while details are difficult to trace today, they seem to have shared a number of similar fundamental ideas about cosmology and religion. In some ways, then, the frontiers of Mesoamerica are not quite so well defined as Kirchhoff and others originally supposed.*

Today we neatly and simplistically divide three thousand or so years of Mesoamerican cultural development and history into three sequential eras or periods: the Preclassic (lasting from about 1200 BC to AD 200), the Classic (AD 200 to 900), and the Postclassic (AD 900 to 1519). These three terms have been around for decades, but as we learn more about the archaeology of the region, the more they seem inadequate and even arbitrary. *Classic* conveys an overly simplistic picture, suggesting that cultural development followed some type of bell curve, reaching an apogee after a long period of development, followed by a quick decline and conquest. The reality is, of course, different and far more complex, and it varies greatly from region to region. Mesoamerican history is perhaps best understood as consisting of the rise and fall of numerous subcultures and cities over time, some connected to one another by language and ethnicity, others not. Generally speaking, the Aztecs, Maya, and other groups never shared any collective identity or sense of ethnicity. Much the opposite was true, in fact, with communities and groups defined in numerous different, fractured, and ever-changing ways. Villages fought villages, cities warred with their neighbors, and occasionally larger territorial hegemonies were forged through conflict and conquest. It's important to realize that ethnic strife and distrust were commonplace, and the cycles of violence must have had a profound effect on the history of the kingdoms, not to mention on the lives of their general populations. As a reminder of

---

* See Wilcox, 1991, for a review of evidence of the Mesoamerican ball game in the American Southwest. Milner, 2004, offers a good overview of so-called Moundbuilder cultures of the American Southeast and Midwest. The brilliant art of Hopewell and Mississippian cultures is presented in Townsend, 2004, by far the best visual source on these unsung cultures of American prehistory.

the power of native ethnic conflict, we need only recall how Cortés was able to defeat the Aztec armies through his powerful alliance with their enemies, who were more than willing—if shortsighted in doing so—to aid the foreigners in the quest to overthrow their rivals.

I find it fascinating that these groups, almost constantly at war with one another, somehow, and for reasons that are still not too well understood, expressed through their art, writing, architecture, religion, and timekeeping a common understanding of the world and humanity's role in it. It's the commonalities among these groups—not their differences—that we will use to define and understand ancient Mesoamerica in all its glory.

A Sketch of Mesoamerican History

Considered by some to be the "mother culture" of Mesoamerica, the Olmec have long been mysterious, and a subject of much debate among archaeologists. Their first impressive monumental remains date to about 1200–1300 BC, mostly at numerous sites near the Isthmus of Tehuantepec, which lies on the narrow portion of the Central American landmass west of Yucatán. Before this time, the inhabitants of this part of the world organized themselves in large and small villages, subsisted on a variety of important agricultural crops, and made use of numerous natural resources in developing trade, craft specialization, and increasingly complex political and religious systems. Centuries of such development laid the groundwork for an important transformation that took place around 1200 BC, when certain villages in the Tehuantepec region and elsewhere reached a certain "tipping point" in their social evolution, perhaps driven by population increases and more efficient agricultural technologies. Village leaders and chiefs now became something approaching kings, able to wield considerable power over their populations, and commissioning monumental artworks and impressive ceremonial architecture. At this early date, we find supernatural and cosmological symbols intimately linked to the institution of rulership, especially at the Olmec site of San Lorenzo, built on a raised portion of land amid lush swamps and lagoons not far from the Gulf of Mexico. San Lorenzo was among the first great ritual and political centers anywhere in Mesoamerica, and according to some, it was the origin point for much of Preclassic royal ideology and religion.

Many Preclassic cultures of the region shared the distinctive art tradition of the Olmec, with its familiar emphasis on rulership and its intimate connection to gods and the cosmos. Evidence of a remarkably coherent and similar iconography, featuring gods of sky, earth, and maize, appeared on monuments and artifacts ranging from the Mexican state of Guerrero to the west all the way to Honduras and El Salvador to the east—generally the same area that we today call Mesoamerica. The surviving artifacts are monumental architecture and sculptural carvings, including famously evocative colossal heads, most likely representing portraits of historical rulers or ancestors.

Other important Olmec sites appeared in other parts of what is now southern Mexico, in Oaxaca and close to the Pacific Coast of Chiapas, near Guatemala. Later sites exhibiting Olmec-style sculpture date as recently as about 500 BC, making for a long era of development lasting at least seven or eight centuries. Today it is difficult to speak of a single "Olmec" culture, for most of the sites of this period are distinct in some way, yet they nevertheless share a remarkably consistent style of art, in monumental and portable sculpture. Rather than seeing the Olmec as one people or one civilization, it is probably more accurate to see the similar remains of this time as a reflection of an initial shared "idea" of Mesoamerica, with cohesion and a codification of art, religion, and political ideology.*

It's probable that the Olmec, or someone living in Olmec times, invented the elements of the calendar we will be examining so closely in this book. We lack any direct evidence of Olmec dates in written form, but the widespread use of, say, the 260-day calendar throughout all of Mesoamerica suggests that it is very old and that the original "mother

---

* For excellent overviews of Olmec culture and the Preclassic era of Mesoamerica, see Diehl, 2004, and Pool, 2007.

culture" may have had something to do with its origin. We do know that writing of some sort was in place in Mesoamerica by about 800 or 700 BC, in later Olmec times, as shown by the recent discovery of the enigmatic "Cascajal Block," a small greenstone object bearing incised hieroglyphs that still resists decipherment.[2] Perhaps the various cycles that make up Mesoamerican time were developed all at once by some brilliant mathematician and daykeeper who lived near San Lorenzo or Tres Zapotes back in the remote days of the early Preclassic period. Until we find some direct evidence of its use by the Olmec, we will never be sure.

During the middle and late Preclassic periods (700 BC–AD 200), Zapotec civilization emerged in the large Valley of Oaxaca, high in the arid mountains of southwest Mexico. The earlier Olmec exerted some influence on the art and religion of the Zapotec, who, along with the Maya, were among the first to develop a distinct regional subculture and sense of local tradition within Mesoamerica. The hilltop ceremonial center of Monte Alban, near the modern city of Oaxaca, was the center of Zapotec social, political, and economic life. Its plazas and pyramids were built over the course of several centuries, and were governed by generations of militaristic rulers who oversaw a complex state that encompassed large areas of the region, including the vast Valley of Oaxaca and beyond.*

The art of Monte Alban reveals that warfare was at the center of the ideology of Zapotec kingship, most clearly seen on the numerous relief carvings showing images of tortured and abused captives. Many intriguing hieroglyphic inscriptions at Monte Alban still remain undeciphered, but we have little doubt that they are written in one of the ancestral forms of the Zapotec language still spoken throughout much of the region. Many of these inscribed tablets contain recognizable dates, recorded using the 260-day calendar so widespread among all Mesoamerican cultures. Even in texts considered unreadable, the date signs are often clearly visible, indicated by their round cartouches and bar-and-dot numbers below.

It's been widely claimed that the Zapotecs were the inventors of writ-

---

* See Marcus and Flannery, 1996, for an excellent and accessible study of Zapotec civilization.

ing in Mesoamerica, although recent evidence from outside of Oaxaca casts some doubt on this long-held idea. An early sculpture excavated at the site of San José Mogote, a few kilometers from Monte Alban, by archaeologists Kent Flannery and Joyce Marcus, shows the image of a sacrificed captive along with a single hieroglyph they interpreted as a calendar hieroglyph, all said to date to sometime before 500 BC. The dating of the stone remains a bit controversial—some think it just might be a little bit later—but at San José Mogote we certainly have *one of* the earliest inscribed monuments in all of Mesoamerica. Some new finds from far away in the Maya region suggest that hieroglyphic writing and calendrics originated there by about 300 BC, suggesting that no one region of Mesoamerica can be singled out as "the" place for the origin of writing.

Ancient Zapotec culture spanned the entire Preclassic, Classic, and Postclassic periods, although in the centuries leading up to the arrival of the Spanish, we see evidence of some political collapse at Monte Alban and an ensuing decentralization of authority in the Valley of Oaxaca. Incursions by neighboring groups called the Mixtecs in the eleventh and twelfth centuries probably contributed to this transformation, and eventually the Zapotec areas of the valley were absorbed into the Aztec tributary empire under the reign of Motecuhzoma II, in the last decades leading up to the Conquest. It was here that Hernán Cortés established his own personal landholdings in 1521, having been granted the title Marques del Valle by the king of Spain.

Moving north from the Valley of Oaxaca, we turn next to perhaps the most interesting yet mysterious place in all of Mesoamerica, the great city of Teotihuacan. Covering a large area on an open plain to the northeast of modern Mexico City, Teotihuacan was at once a city and a civilization. Between roughly 100 BC and AD 600 it was the single greatest urban area in all of the Americas, dominating much of Mesoamerican commerce and politics and serving as an important religious center. At the time of its demise, it was among the largest cities in the world: the population of the city and its immediate surroundings near the Valley of Mexico was estimated to have been as high as 200,000. Teotihuacan's importance in broader Mesoamerica is clearly indicated by its extensive trade networks and strong political and ar-

The Temple of the Sun at ancient Teotihuacan, the highland metropolis that dominated much of Mesoamerica from 100 BC–AD 500. (Photograph by G. Stuart)

tistic influences found by archaeologists nearly everywhere in Mesoamerica, from as far away as Oaxaca to the south and the Maya region to the east. Its widespread importance and power seem undeniable, but despite leaving such a heavy mark on the region's archaeology, Teotihuacan remains an utter mystery in so many respects. We have no firm idea of its social structure, its government, the history of its rulers, the language its people spoke, or why its influence was so profound.*

The city's broad layout conformed to a careful gridlike plan—highly unusual for a Mesoamerican city—with numerous city blocks, apartment complexes, and channeled rivers. All of it was oriented along a wide, grand avenue nearly a mile in length, known today as the Street of the Dead. Near the street's northern terminus are the pyramids of the Sun and the Moon—among the largest buildings ever constructed anywhere in the world at the time. (Their names, given centuries later,

* See Berrin and Pasztory, 1993, for a general look at the art and archaeology of Teotihuacan, which includes essays from numerous scholars who have worked there over the years.

are not original.) These massive buildings were man-made mountains, and later Aztecs were amazed to think they were built by an earlier people so remote in time. The larger of the two buildings, the Pyramid of the Sun, is one of the biggest constructions in the ancient world, built in the second century AD above an artificially modified cave—a ritual locale that, like many other caves in Mesoamerica, was perhaps considered a place of human origin from the earth. We will never know the specific meanings of many of these structures.

During the fourth and fifth centuries, for reasons still unclear, Teotihuacan's rulers and elites had a profound interest in the distant lowland Maya. We see evidence of this in both the written history and the material culture from a select number of Maya cities, where monuments in the city centers depict foreign warriors who seem to have had a profound impact on local Maya dynasties. The most significant of these figures was perhaps Siyaj K'ahk', a "western" lord who, according to the detailed written history left by the Maya, journeyed toward the heart of the Maya region and established some sort of political authority near the important kingdoms of Tikal and Uaxactun. Another important figure in the story of Teotihuacan-Maya interactions was a ruler named K'inich Yax K'uk' Mo' (Great Sun Green Quetzal-Macaw), the founder of the remarkable dynasty centered at Copan, in present-day Honduras.[3]

I doubt we will ever know exactly who built Teotihuacan. We have no firm idea even what language the people there spoke, and details of its political history are forever lost, given the complete lack of historical records from the city. But considering that less than 10 percent of the ancient metropolis has been scientifically excavated, I am optimistic that further explorations will reveal many surprising things. Only a few years ago, excavations in the large pyramids of Teotihuacan led to the discovery of royal tombs within one of the great pyramids—one of the first good indications that divine rulership played a role in the city's political and social fabric.[4] This may appear unsurprising given the neighboring cultures of Mesoamerica, but the lack of many ruler portraits and historical texts has led some to propose that Teotihuacan had a radically different political structure than its Mesoamerican counterparts.

The Aztecs, living a thousand years later, were in awe of the vast ruins northeast of their capital. They rightly claimed to have inherited from Teotihuacan's ancient inhabitants many of their own religious beliefs and cultural institutions. Teotihuacan also held mystical significance for the Aztecs as a place of world creation, where the gods themselves were believed to have first manifested. The Aztec legends of successive creations and destructions, to be discussed at length in chapter 7, included an account of the beginning of our current era, called the Fifth Sun, when many of the would-be gods of the cosmos gathered at Teotihuacan in darkness around a sacred fire. One of the participants was the humble Nanahuatzin, who cast his corporeal self into the flames of the hearth and transformed into the sun, Tonatiuh. The others gathered around the fire likewise sacrificed themselves to become the gods of the Aztec pantheon. It comes as no surprise that in much later historical times, Aztec rulers made regular pilgrimages to Teotihuacan to pay homage to Nanahuatzin and his companion gods; it's difficult to overestimate the importance of Teotihuacan to Mesoamerican history and culture.

Mesoamerica as a whole seems to have been greatly affected by the downfall of Teotihuacan in the late Classic period, roughly between AD 500 and 600. The once-great power of religion and politics met a violent end and was soon abandoned, its population scattered to other areas of the Valley of Mexico. In the slow wake of this demographic and cultural turmoil, new political centers rose and fell, some adhering to familiar patterns of interaction with other parts of Mesoamerica, some emphasizing long-held cults of militarism and religious symbols, such as the feathered serpent Quetzalcoatl. Even after its demise, Teotihuacan left behind a powerful legacy that influenced all later cultures of highland Mexico, all the way up to the time of the Spanish Conquest.

One of the most important highland states of this time was centered at Xochicalco, an impressive hilltop ruin located not far from the modern city of Cuernavaca, in the state of Morelos. Its origins can be traced back to the Preclassic era, but the vast amount of what one sees there today—courtyards, pyramids, ball courts, and sculpted monuments—dates to the late Classic, between approximately AD 700 and 800. The buildings are nowhere near the overwhelming scale we see at Teotihuacan, but Xochicalco is nevertheless among the largest sites of its era. The most prominent building is what archaeologists call the Temple of Quetzalcoatl. It is a large masonry platform sculpted on nearly its entire surface with designs of undulating feathered snakes intermingled with images of seated rulers who have a distinctively Maya look to them. Among the imagery are examples of large hieroglyphs depicting calendar dates that are probably historical or mythological references of some sort. This "hybrid" culture, combining Maya and Mexican elements, is reflected in the site of Cacaxtla, with its spectacular murals depicting the defeat of a Maya-looking army by what are presumably local warriors.*

Militarism continued to be the overarching theme of these central Mexican cultures, but none emphasized it more than the site of Tula, located in what is today the state of Hidalgo, north of Mexico City. Tula, or Tollan, was the center of a prominent "Toltec" city that rose to prominence a bit later than Xochicalco, reaching its height as a political center between about AD 900 and 1100. Its influence is strongly felt in far-off Yucatán, at the ruins of Chichen Itzá, where Toltec styles of art and architecture intermingle with Maya forms. This intimate connection between Tula and a distant culture to the east may follow a general pattern earlier seen with Teotihuacan and the Maya, but archaeologists still debate whether the rulers of Tula conquered or otherwise established a major ritual center in distant Yucatán.

The early Postclassic period—the two to three centuries after AD 1000—seems to have been a particularly unstable time throughout Mesoamerica. Tula's collapse, like that of many others, is not at all well understood, but it clearly left a major vacuum in the power structure of the Mexican highlands in the thirteenth and fourteenth centuries. This looks to have been an era of balkanized polities and rival city-states waiting for the next cycle of political and military consolidation. These years saw the dramatic rise of several important polities in the highlands of Oaxaca, ruled by Mixtec peoples, who, as noted,

* For a comprehensive treatment of the archaeology of Xochicalco, see Hirth, 2000. For the murals of Cacaxtla, one widely accessible source is G. Stuart, 1992.

seem to have taken advantage of the waning centralized authority of the Zapotecs, who had ruled much of Oaxaca for centuries.

In the basin of Mexico, the stage was now set for the arrival and eventual expansion of the Aztecs, or Mexica, as they called themselves, the last great civilization to rule in Mesoamerica. By the early sixteenth century, on the cusp of Cortés's arrival, the Aztecs controlled a large part of the territory that defines present-day Mexico, ruling from their island capital of Tenochtitlan, now beneath Mexico City. It was their empire's highly centralized political and economic structure that was turned to such great advantage by the Spanish after 1521. With the conquest of Tenochtitlan, the Europeans already had an infrastructure in place by which to rule and expand their hegemony in New Spain.

The Aztecs' sudden rise to power in the region, at least as their own historians told it, is one of the great rags-to-riches transformations in all of human history. According to the semi-mythical accounts left to us in the form of written translations by Spanish priests, the Mexica were one of a group of nomadic tribes that wandered throughout much of western Mexico in the remote past, making their way eventually to the Valley of Mexico and the fertile area of Lake Texcoco. It is difficult to parse history from myth in those native accounts, but we can say that the first Mexica rulers established Tenochtitlan as a political and ritual center in AD 1325 (the year 2 House in the Aztec calendar); now the stage was set for a long series of conquests and alliances. From that point, the empire expanded rapidly, especially under the conquests of the emperors Motecuhzoma I (1440–1469) and Ahuitzotl (1486–1502).

Most of our extensive knowledge about Aztec culture and religion derives from the remarkably detailed accounts collected by the Spanish in the sixteenth century, when European and pre-Columbian worlds coexisted for a time. We know from such accounts that the Aztec ruler Motecuhzoma II—familiar to most as "Montezuma"—seems to have been a pious fellow, ever conscious of his religious duties as supreme ruler, or *huey tlahtoani*. If we believe the historical sources written shortly after his death, he was a devoted king who paid honor to the

Indigenous map showing the foundation of the Mexica (Aztec) capital at Tenochtitlan, symbolized by the eagle atop the cactus; from the sixteenth-century Mendoza Codex.

myriad gods housed in the temples of Tenochtitlan, his capital, and who carefully tended to the numerous important rituals required of him as leader of the Mexica-Aztec world. Many of these duties were based on the premise that he was a divine being among humans, an integral force within the greater cosmos just as much as the sun, the moon, the mountains, and the earth. Like so many Mesoamerican rulers before him, he resembled a high priest, as he presided over myriad public ceremonies around Tenochtitlan and environs. He was not a priest in the true sense—kings probably knew little of the intricacies of the sacred text, the *tonalpohualli;* nor did they interpret books of prophecy—but as a ceremonial actor, the incarnate sun, he engaged with gods, sanctified objects through his presence, and renewed the world.

Perhaps the greatest religious event in Motecuhzoma II's reign was the so-called New Fire ceremony of 1507, properly known as Xiuhmolpilli, or the "Binding of the Years." This ceremony was the time for world renewal, which was held every fifty-two years. It was a time when the fires of the world were relit atop a small sacred mountain named Huixachtlan, or "Beside the Huixache Tree" (an acacia). The ceremonial drilling of the new fire, created by rubbing two sticks together quickly to ignite a flame, was a reenactment of the sun's birth from a divine turquoise hearth. As the great work of the Spanish Fray Bernardino de Sahagún, the noted chronicler of Aztec culture and history, describes it:

> First they put out fires everywhere in the country round . . .
> And when it came to pass that night fell, all were frightened and
> filled with dread. Thus was it said: it was claimed that if fire
> could not be drawn, then (the sun) would be destroyed forever;
> all would be ended; there would evermore be night. Nevermore
> would the sun come forth.[5]

Atop the hill the priests watched the night sky, paying close attention to the movement of the constellation of Pleiades, called the "fire drill" by the Aztecs. Once it had reached the zenith point, directly above them in the sky, the New Fire ceremony began. The fire was drilled by priests in the open chest cavity of a prone sacrificial victim, whose heart had been extracted by the priests who'd accompanied the emperor to the mountain. Thus the heat source of the world served as a symbolic substitute for a human heart, the source of life and heat that guaranteed that the world would not end. Once lit, the fire was taken down from the hill and distributed among all of the people in Tenochtitlan and around Lake Texcoco.

In antiquity, this all-important hill was perhaps the most sacred of points in the ceremonial landscape around Tenochtitlan. Today, it is more an island in a sea of concrete buildings and busy streets, seldom visited by tourists and consumed by the sprawl of modern Mexico City. Recent archaeological investigations there suggest that the ceremonial importance of Huixachtlan hill predates the Aztecs by many centuries, as suggested by the impressive temple remains from earlier Teotihuacan times. Also, there are good indications that such world-renewal ceremonies were once more geographically widespread throughout central Mexico, but that they became more centralized around Tenochtitlan and its environs with the rise of the Aztec Empire.*

Motecuhzoma II's actual role in the New Fire ceremony is not very clear from the descriptions left to us—his "fire priests" were doing most of the hard work igniting the sticks, it appears—but in the state religion of the Aztecs, he would certainly have claimed credit and responsibility for its results. Perhaps his most important act in connection with the great ceremony was to instigate a change in its timing. On earlier occasions, the New Fire ceremony was always in the year 1 Rabbit, exemplified in the recording of the New Fire ceremony in 1454. But that year was infamous in Aztec historical memory as a time of great famine, and it so happened that the following 1 Rabbit, in 1506, was shaping up to be a repeat of the same terrible drought in the Valley of Mexico. Ever superstitious, Motecuhzoma II opted to avoid a world

---

* See Elson and Smith, 2003, for archaeological evidence of the New Fire renewal ceremonies in widespread parts of central Mexico.

renewal under such traumatic circumstances. As the supreme ruler, he simply changed the calendar, making the following year 2 Reed, 1507, a "new and improved" time for the New Fire.*

A fascinating sculpture in Mexico City provides some indication of just how important the New Fire ritual was to Motecuhzoma II. In the "Central Park" of Mexico City, at the base of Chapultepec Hill, carved into a natural rock outcropping, are the meager remains of Motecuhzoma II's "official portrait." This once-elegant sculpture is now barely discernable to the locals and tourists who stroll by it every day. The standing figure was carved in 1519, just months before the tumultuous events that would bring a quick end to the Aztec world. A few short years later, the Spanish intentionally obliterated the carving, leaving it in the condition one sees it today.

The hieroglyph identifying the portrait as that of Motecuhzoma II is just visible next to his body, as is another glyph, reading, "2 Reed." Although it was carved a dozen years after the event, the Chapultepec portrait highlights the reformed timing of the ceremony, suggesting that it was one of the emperor's great and proud achievements. How ironic, then, that within months of this commemorative portrait being carved, Motecuhzoma II succumbed to a foreign foe, a defeat that brought about, in a very real sense, the end of a great era. No New Fire could have forestalled that.

Motecuhzoma II's close attention to performance and ceremony grew in no small part out of his ideological role as the earthly embodiment of the sun. In the pageantry of many rituals, Aztec emperors routinely assumed the mantle of the sun and its power, often as

---

* Such royally decreed calendar reforms or tweaks may have been more frequent in Mesoamerican history than we've previously thought, but direct evidence for this is hard to find. I suspect that the highly centralized infrastructure of Aztec politics and religion made it easier to dictate such reforms than, say, was the case among the classic-era Maya, where many kingdoms shared a single calendar system. No one Mayan lord could simply choose to modify a period-ending celebration based on a bad prognostication, since all Mayan kings, rivals, and allies had to celebrate many of the same rituals at the very same time.

The defaced portrait of the Aztec emperor Motecuhzoma II, carved on a stone outcrop of Chapultepec hill in Mexico City. (Photograph by the author)

impersonators of solar deities such as Huitzilopochtli and Tonatiuh. The Spanish priest and historian Diego Durán stated that the early Aztecs likened the death of a ruler to the onset of a solar eclipse, and the subsequent installation of a new ruler, or *huey tlahtoani*, to the sun's reappearance after darkness. Durán mentions that the accession of the young Ahuitzotl, Motecuhzoma II's predecessor and uncle, brought a new cosmic light onto the world: "When the boy had been thus enthroned. Tlacaelel (the royal advisor) sent messages to the different regions to proclaim to all that the sun, which had been darkened, once again shone in Mexico."* World history shows that absolute royal authority can easily spawn metaphors and ideas of "sun kingship," and I'm willing to bet that Ramses II or Louis XIV, Europe's great Sun King, would probably have admired and maybe even identified with many of Motecuhzoma II's emblems of status and power.

It's fascinating to read these accounts of the ruler performing so central a role in cosmic renewal ceremonies, knowing that Motecuhzoma II was apparently barely cognizant of the arcane details of the calendar and its interpretation. Kings were cosmic actors who often embodied time, yet they left the meaningful study and analysis of time's structure up to astrologers and soothsayers. The irony comes through in the historical accounts of Motecuhzoma II's anxiety over a series of troubling omens during the years and months leading up to the Conquest. In 1517, a large comet suddenly appeared in the night sky (like "a wound in the sky, wide at its base and narrow at its peak"). In addition, there was a spontaneous fire in the Temple of Huitzilopochtli, the Mexican Aztec tribal god, and a particularly bright meteor streaking across the sky. These and other omens were said to have caused great angst and terror among the Aztec people. Later news of the appearance of strange "floating houses" along the coast spread throughout the empire—and information did travel surprisingly fast by foot in those days—yet another bad harbinger. Motecuhzoma II was keen to under-

* See Durán, 1964, p. 185. For a discussion of solar metaphors involving Aztec kingship, see Umburger, 1986.

stand the significance of what was happening. He called in calendar priests from many different towns, eager to know what the prophecies in their books might tell him of his fate. "What can we say?" they answered, not having seen the same fearsome omens. All they could muster was a vague prediction that "Motecuhzoma will behold and suffer a great mystery," but nothing more. The soothsayers escaped from the clutches of the emperor that night; in retaliation, the frustrated king had the priests' families massacred.[6]

Motecuhzoma II, like all Mesoamerican rulers, therefore had a complex and even at times contradictory relationship with cosmic time, playing the role of both giver and receiver of cosmic power and influence. If we didn't have the intimate Spanish historical accounts telling of his angst and his inability to foretell the future, we would see only what the official Aztec art and texts had to say: they incessantly emphasize the more active role of Mesoamerican kings as shepherds of time and its structure. Throughout Maya and Aztec art we see ample reminders that rulers were thought to have a very close, even inseparable identification to time as a cosmic dimension. Motecuhzoma II and other Aztec rulers were believed to have the symbolic ability to renew the world and reenact the birth of the sun.

The Maya

Maya civilization, the focus of this book, rose and developed over the course of many centuries, and spanned most of the course of Mesoamerican history. The origins of its famous art style and monumental architecture can be traced far back to sites of the Middle Preclassic period, if not before, and over the centuries, through numerous rises and falls, its culture endured up to the time of the Conquest. In fact, Maya communities along the east coast of the Yucatán Peninsula were the very first Mesoamerican cities to impress Spanish sailors in the first decade of the sixteenth century, not long before Cortés's first journey into the heart of Mexico. Geographically speaking, the Maya occupied a very large area that is far more topographically varied than that of other Mesoamerican cultures we've discussed. This region encompasses nearly all of Mesoamerica to the east of the Isthmus of Tehuantepec, from the scrublands of northern Yucatán to the high rain forests of the central Petén, and up into the cool highlands of what is now Chiapas and Guatemala.

The meaning of the term *Maya* has changed significantly over the years, and even today its use needs to be carefully considered and explained. Originally, from the time of the Conquest up to the twentieth century, the word applied only to the natives of the northern part of Yucatán, speakers of what we now call the Yukatek (or Yucatec) language. This area was historically known as Mayab before the arrival of the Spanish, and *Maya* was an ethnic term used by the inhabitants in reference to the larger cultural area, incorporating a number of native states tied together by language, economics, and culture. Significantly, in this time no indigenous groups outside of Yucatán ever used the word *Maya* in reference to themselves. It's significant, too, that Yucatán (Mayab) was the region where a unified approach to the ancient and modern natives took hold, one that acknowledged that the indigenous populations of Yucatán were the direct descendants of the builders of nearby archaeological remains in the area—an argument forcefully advanced by explorer John Lloyd Stephens as early as 1841. *Maya* thus took on an archaeological meaning; "Maya ruins" were seen everywhere, and explorations throughout the more remote forests of Chiapas and Guatemala revealed even more "Maya" remains. Early scholars of indigenous languages also recognized that the Mayan language spoken in Yucatán was related to K'iché', Tzotzil, Tzeltal, and other languages found in the mountains to the south. In time, these came to be grouped together as "Mayance" or "Mayan" languages, giving priority to the ethnic term that had before been applied only to the natives of Yucatán. It wasn't long before *Maya* and *Mayan* came to be widely applied by anthropologists and historians to the many different peoples and languages of the area, and to those who never even made use of the terms in referring to themselves. Even today, relatively few K'iché' or Tzotzil people would prefer to call themselves Maya, given that the term has little ethnic meaning or frame of reference for them.

Maya has become an ethnic term largely imposed on diverse but related groups from outside forces and the media. Today, it has taken on a new significance as native groups have forged new collective identities, and as indigenous political and social activists tap in to a rich archaeological legacy, hoping to create a more powerful coalition of indigenous groups who had been far too diverse.

Let's look at what we know of the archaeology of the Maya. By 500–400 BC, around the same time that the Zapotecs were just forming their militaristic state in the Valley of Oaxaca, Maya civilization was also taking shape at sites throughout Yucatán and the Petén. Large villages grew in population, giving rise to new social and political institutions, including a powerful ideology of divine kingship. In larger towns, carefully planned monumental architecture was built using massive amounts of human labor. The earliest buildings of sites such as Nakbe, El Mirador, Naranjo, Tikal, San Bartolo, and other great centers date to this middle Preclassic era, all reflecting an extraordinarily unified "look and feel," with their ambitious design and use of modeled stucco decoration. The terraces and platforms of most pyramids at this time were adorned with beautiful "masks" of gods and ancestors, no doubt symbolizing the divine connections of early Maya kings. By 200 BC, numerous centers of political authority were firmly on the Maya landscape, from northern Yucatán (at Acanceh, for example), to the Petén in the south (at sites such as Chakanbakan, El Mirador, and Uaxactun).

Some of these early Maya centers are truly extraordinary in size, dwarfing much of the later architecture from the Classic era. At El Mirador, for example, the main pyramids, called Danta and El Tigre, are among the largest buildings of all of Mesoamerica. Major networks of roadways connecting these and nearby sites in the so-called Mirador Basin (not really a basin) can still be seen from the air, crossing the low-lying areas that may once have been seasonal lakes. Today there is virtually no water in this remote zone, but around 400 BC it must have been far wetter and able to support tens of thousands of people in the surrounding territory. It was probably a profound environmental change that led to the abandonment of Mirador and surrounding centers by AD 200, when the nearby centers of Tikal and Uaxactun gained in population and political importance. These rises and falls of regional centers were typical of the Maya political landscape throughout history, up to the arrival of Europeans centuries later.*

One of the most extraordinary of all Preclassic Maya sites is San Bartolo, recently discovered in the remote part of northern Guatemala, not far from the ancient cities of Tikal and Uaxactun. In 2001, my friend and colleague Bill Saturno first visited San Bartolo's unknown ruins and came across evidence of recent looting. (Tragically, organized looting has long been destroying archaeological remains in the area.) Inside a large excavation tunnel in one of the largest buildings, Bill was amazed to find a small exposed portion of a painted wall, with fine line work and rich colors of red, cream, and yellow. Maya wall paintings are rare from any time period, but these were clearly Preclassic in style, and therefore almost unique in all of Maya archaeology.

Since Bill and I were both working at Harvard's Peabody Museum of Archaeology and Ethnology, he alerted me at once to this important find. We later returned to the ruins with a group of conservators to assess and try to protect the fragile mural. After we worked closely with the Guatemalan government, the San Bartolo Archaeological Project began officially, and Bill began his excavation project in Las Pinturas (The Paintings), as the ancient pyramid was soon called. The digs in the buried chamber first found by the looters revealed forty-six feet of decorated wall still in place, left over from a building that had been intentionally destroyed by the Maya when they refurbished it in around 100 BC.

Now visible on the walls and in tens of thousands of fragments scattered on the ancient floor of the pyramid, the San Bartolo murals have taken a rightful place as one of the great artworks of the ancient Americas. From what we can see at these initial stages (documenting and studying all of the fragments are still under way), its scenes

---

* These and other developments in archaeology are nicely summarized in Martin and Grube, 2000; Coe, 2005; Sharer and Traxler, 2006; and Houston and Inomata, 2009.

A scene from the early Maya wall paintings at San Bartolo, Guatemala, showing a young lord performing a bloodletting ritual. (Photograph by the author)

are purely mythological and seem to relate episodes from the story of Maya Creation. We have found at least four discrete sections of the overall painting, each with a different theme. One concerns the emergence of the maize god from a sacred mountain of sustenance; another involves the positioning of four "world trees" in the quarters of the cosmos and their associated sacrifices of animals and human blood— a theme that is known from much of Mesoamerican cosmology. Another scene shows a prototypical Maya king being crowned as he sits upon a modest-looking scaffold throne. More pieces of the San Bartolo story are now being fitted together, literally, and with each piece of the puzzle found, our understanding of early Maya religion and cosmology changes. One clear result of this mural discovery is the evidence that the Preclassic Maya were anything but undeveloped in terms of art and ideology. By 200–100 BC, Maya civilization of the lowlands was going very strong; evidence in the mural's style, lyricism, complexity, and use of color and tone indicate this visually. In many respects, what archaeologists once thought of as exclusively "Classic Maya" has turned out to have complex antecedents in the earlier Preclassic.*

With the arrival of the so-called Classic period, spanning roughly AD 250 to 850, we have a much clearer picture of Maya civilization's inner workings, as a result of detailed archaeological and historical evidence. For these six centuries or so, great city-states and dynasties reemerged in the southern lowlands, often in conflict, but with each participating in a well-defined civilization where kingship and regal ritual took center stage. Prominent centers such as Palenque, Copan, Calakmul, and Tikal were the focal points of complex political alliances and courtly intrigues, where kings and their families, as elsewhere in world history, vied for power, prestige, and religious favors, while at the same time dealing with many real-world problems of politics and

---

* For a brief overview of the San Bartolo excavations, see Saturno, 2003 and 2006. A more technical account of the interpretation of a portion of the mural is found in Saturno, Taube, and Stuart, 2005. The complete analysis of the mural and its iconography is in preparation.

economics. In Maya history there were several dozen of these cities, and they varied considerably in their population, size, and influence. Each dynastic center also had its own distinctive sub-culture that, while still Maya, employed subtle differences in styles of art and architecture.

Maya Writing

One of the great keys to understanding Classic Maya history and religion is the remarkable hieroglyphic writing system used by the Maya for some two thousand years for composing official texts found in temples, on monuments, and on various portable objects. Maya court society was a literate world, where the written word assumed a sometimes sacred role, even at times recording the actual spoken words of kings and ancestors.

The breaking of the Maya "code" has a long and colorful history involving a number of interesting characters. By the 1880s, Ernst Förstemann in Germany and Joseph Goodman in the United States had deciphered the hieroglyphs that explained the Maya calendar, using one of the four known original Maya books to have survived the Spanish Conquest. (We will look at their accomplishments in some detail in chapter 6.) In the wake of their key insights, Maya archaeology became largely obsessed with dates and chronology, and even led a few myopic scholars of the time to claim that the Maya had little else to write about *except* dates. By the 1930s and '40s, most Maya glyph research was concerned primarily with tracking down and studying the dates written on stone monuments, with little regard to what the dates themselves might actually signify. For early generations of Maya archaeologists, the ancient Maya were brilliant timekeepers, but were seemingly unconcerned with the humdrum world of history, governance, and everyday life.

This all changed in 1960, with a humble proposal from a museum curator named Tatiana Proskouriakoff. Her family had immigrated to America from Siberia in the early 1900s, and by the 1930s Proskouriakoff became one of the select women of her time to earn a university

Two Maya ruins of the Classic period: Tikal (top) and Palenque (bottom).
(Photographs by the author)

Maya hieroglyphs carved on a monument from Dos Pilas, Guatemala, recording royal history of the seventh and eighth centuries. (Photograph by the author)

degree, specializing in archaeological illustration. After an early career as a rare female artist and scholar working on male-dominated archaeological digs in the Maya region, she was hired by Harvard's Peabody Museum of Archaeology and Ethnology as a Research Fellow in Maya Art. There she was free to direct her own research, using the extraordinary Maya art collections found by earlier explorers, hauled from Mesoamerica to Cambridge, and housed in the museum's basement. Years earlier, Proskouriakoff had participated in the field research at the ruins of Piedras Negras, Guatemala, famous for its beautifully carved limestone monuments. Long after her work on that project, she continued to plug away at deciphering the dates and stylistic evolution of its tall sculptures, or stelae, and by the late 1950s she had made a remarkable breakthrough. Proskouriakoff recognized that most of the forty-odd stelae from Piedras Negras could be grouped together by their motifs into discrete sets, or "series," seven in all. Each of these series began with a monument showing a richly clad man seated on a raised throne (see page 263). Every few years more monuments would be added to the series, some showing similarly dressed protagonists engaged in rituals or, at other times, as warriors standing above captives. To Proskouriakoff, the portraits of these men, often surrounded by waiting attendants or the symbols of military victory, smacked of historical content. Taking this as a cue, she noted that the dates of each series of monuments spanned no longer than a few decades. Furthermore, the inscriptions on these monuments featured certain "event" or "action" glyphs with knowable dates that fell at certain points within the spans of each series. One action glyph was always the earliest of the set, while another distinctive glyph always took place two or three decades after that "initial" event. Proskouriakoff took a not-so-radical leap to propose that the events in these Piedras Negras monuments were episodes of life history: the initial event could reasonably be seen as indicating "birth"; the second event in early adulthood appeared to function as an event of crowning or inauguration. In one fell swoop she had teased out a complete dynastic history of a Maya kingdom. After her paper was published in 1960, Maya studies would never be the same.*

Just a few years before Proskouriakoff's breakthrough, and on the other side of the world, another scholar (coincidentally with a Russian name) was making profound observations about the linguistic content of Maya writing. Yuri Knorosov, a young philologist working in Leningrad (today's St. Petersburg), was fascinated by exotic scripts of the ancient world, and had spent many years studying Egyptian

---

* See Proskouriakoff, 1960, for her analysis of the Piedras Negras "pattern of dates." An excellent biography of Proskouriakoff is Solomon, 2002.

Portrait of the ruler Itzamnaah K'awiil, who reigned over the site known today as Dos Pilas in the eighth century AD. He wears the mask of a rain deity as part of a ritual calendar dance. (Photograph by the author)

hieroglyphs and Sumerian cuneiform. Knorosov had come across published copies of three of the four remaining Maya books; named for the cities in which they were found, they are now called the Dresden, Madrid, and Paris codices. He found these copies among shipments of scholarly books that had arrived in the Soviet Union as loot after the Battle of Berlin in the Second World War. Knorosov dove into these bizarre-looking pages, and looking at the glyphs, he was able to discern some familiar patterns. He could make a reasonable case, he felt, that the signs of Maya writing are not all word signs (as scholars in Europe and the United States had long emphasized), but rather syllabic, spelling words in a fully phonetic manner. He published his conclusions beginning in 1952—at the height of the cold war. The Soviets had proudly championed Knorosov's use of "Marxist-Leninist methods" to develop his theory, all of which predictably led to stinging critiques from the likes of J. Eric S. Thompson and other prominent Mayanists of the West. It took decades for Knorosov's fundamentally true insights to gain acceptance among a wider array of Mayanists, but by the mid-1970s his methods were considered generally sound.

Much tweaking of the rules and approaches used in deciphering hieroglyphic writing took place throughout the 1980s. Over the last twenty years or so it is fair to say that we have deciphered the language of Maya hieroglyphs. While Maya script is visually complex, its underlying principles are fairly simple. It is a fully phonetic writing system, making use of two kinds of signs or characters: word signs (or logograms) and consonant-vowel (CV) syllables, such as *ma, ʃi,* or *lu.* A good example of how Maya writing works can be illustrated by the spelling of one of the months used in the solar year of 365 days. The month named *Sotz'* or *Suutz'* (depending on which Mayan language you use) means "Bat," and was used as the proper name of the fourth month of the solar year calendar. Usually this name was "spelled" simply with the profile of a leaf-nosed bat's head—a direct picture to represent the name. On rare occasions, the word *Suutz'* could be directly spelled in a purely phonetic manner using the combined syllables *ʃu* and *tz'i.* Either method was fully acceptable, and both might even be employed by the same scribe in the same piece of writing. The complexity of Maya writing was greatly enhanced by the given scribe's love of variation. Any given syllable, such as *ma,* could have several different visual forms, all perfectly equivalent and acceptable in the spelling of words. If scribes opted to use such variations, it became difficult for a modern scholar to discern just what the ancient writer was trying to say. For this reason more than anything, it took decades for scholars from all over the world (including me) to figure out the visual structure underlying Maya writing. After years of collaboration, meetings, scholarship, and research, we finally figured it out in the late 1980s; it was only then that we could consider Maya glyphs to be "deciphered," more or less. Now, thanks to the work of so many scholars, it's possible to read about 80 to 90 percent of all Maya texts. Not bad, considering that virtually nothing was readable half a century or so ago.*

---

* Much of the history behind the deciphering of Mayan glyphs is in Coe, 1992. For a more detailed look at how Mayan glyphs operate and are constructed, see Coe and van Stone, 2001; and Montgomery, 2002.

The deciphering of Maya glyphs has moved Maya archaeology and research away from its non-historical roots in American archaeology and toward something more like Egyptology and classical studies. Now the written record can be studied in close conjunction with traditional archaeological evidence. Not long ago, scholars thought the Maya were "mysterious" and believed they had no written history; I find this very strange in retrospect. Grounded now in history, and with a wealth of archaeological artifacts in hand, we can closely study not just Maya kings and queens, but the larger population as a whole, and try to understand just how their civilization evolved and changed (sometimes radically) over the centuries.

One of the great long-standing "mysteries" about the Maya has centered on their apparent collapse, or what has often been erroneously called their "disappearance." As mentioned earlier, the Maya did not disappear, for they are still with us, living and working in Mexico, Guatemala, Belize, Honduras, and El Salvador. But it is certainly true that the end of the Classic period, between about AD 800 and 900, many cities were abandoned, and their supporting populations moved on to other areas. We are still debating how and why this happened, and many explanations have been offered over the years. Was it disease? A revolt of the everyday people over the elite rulers? Was warfare a factor? In fact, we think that many of these components probably played a role in this important transformation of ancient Maya culture, perhaps combining to make continuation of the old ways impossible. My own strong sense is that there was no single factor to explain the collapse of the Classic-era Maya. As we have touched on earlier, one factor may have kicked the process forward: population growth. Archaeological evidence from across the Maya region attests to a rapid growth in numbers of people after about AD 600, and after several generations I can see how such numbers would have made daily life a special challenge, particularly in areas where water and arable agricultural land were already scarce. Warfare and disease are also both well documented in the archaeological record of the time, and can be seen as important ingredients in the complex stew of problems facing the Maya.

As with the end of the Late Preclassic period centuries earlier, the end of the Classic period saw major transformations in the nature of Maya politics and ideology. For reasons still not terribly clear, dynastic rulers were no longer the focus of the monumental art and inscriptions. A number of important centers rose to prominence in northern Yucatán, at cities such as Uxmal, Nohpat, and Kabah, only to fall and be abandoned within several generations, again for reasons hard to discern. In this period of profound demographic change, influences from highland Mexican cultures to the west appear to increase. There, on the plains of Yucatán, a new and very different sort of ceremonial center arose at Chichen Itzá. So-called Toltec styles of art and architecture appear everywhere at Chichen Itzá beginning around AD 1000, but nowhere else in the Maya area. For some reason, Chichen Itzá seems to have been targeted as the center of a foreign settlement from the highlands of central Mexico emphasizing religious and military symbolism quite unlike anything that had existed before. Much of the iconography of Chichen Itzá emphasizes militarism and the religious "cult" of the feathered serpent—what the Maya of Yucatán called K'uk'ulkan and the Aztecs would later call Quetzalcoatl—a complex but still poorly understood deity associated with the planet Venus, the wind, and rulership. The feathered serpent image had a long history in central Mexico at centers such as Teotihuacan and later Cholula, but it is not until AD 1000 that it becomes a major symbol in the Maya area. I suspect that this new center at Chichen Itzá was created in large part because of the site's immense sinkhole, or cenote (*ts'onot*), one of the most prominent natural sacred sites on the entire peninsula. This place of great supernatural significance was probably well known throughout Mesoamerica, and it may have drawn intruders from the highlands who wished to expand militarily and appropriate a major pilgrimage center in the process.

The "Toltec" presence at Chichen Itzá itself did not last long. By about 1200 its importance as a center of political and religious activity had waned, and Yucatán again became an area of numerous small rival but interconnected states. The native Spanish historical sources from the colonial period are somewhat confused and contradictory about the political situation in Yucatán after Chichen Itzá's demise, but the records and the archaeological evidence show that much of the political and ritual power of the peninsula shifted to the walled city of Mayapan, located to the west. Its overthrow in the early fifteenth century set the stage for the fragmented political landscape encountered by the Spanish conquistadors in the 1500s. The earlier end of Chichen Itzá may have spurred the movement of the ruling lineage toward remote Lake Petén Itzá, where generations of leaders, all named Kanek', would continue to rule for hundreds of years.

By the fifteenth century, on the cusp of the Spanish Conquest, numerous independent Maya kingdoms again dotted the landscapes of Yucatán and highland Guatemala. Between them, in the most remote jungles long ago abandoned by the Classic-era Maya, the Itzá kings named Kanek' begun to rule over a large state centered on Nohpeten. They, as we've seen, are the ones who greeted Cortés in 1524 and later welcomed with suspicion various Spanish friars who urged their conversion to the Christian faith. This came finally with the overthrow of the last Kanek' in the spring of 1697.

Ancient Mesoamerica was a diverse cultural terrain densely populated with speakers of hundreds of languages, living in hamlets, villages, and cities over the course of some two thousand years. They resided in marshlands and mangroves, rain forests, arid plains, and under snow-capped volcanoes. And yet out of all that difference and history arose a few common threads regarding their life, history, and worldview—a common perspective on the world and how it operates. This was Mesoamerican civilization. Among the thoughts, lifestyles, and important factors that linked these diverse peoples was a single calendar. It was a way of organizing time and space into a wonderfully cohesive and coherent system that found relevance among all members of society, not just the kings and the temple priests. It's there that we now have to turn, for to really understand Mesoamerica's culture and history, one has to start on time.

3

THE ESSENCE OF SPACE

Hail divine world, however many your manifestations in this earth . . .
—Opening of a K'iché Maya prayer[1]

The new sewing machines began having problems soon after they came to San Lucas, a remote Q'eqchi' Maya village in the uplands of Guatemala. Almost immediately after a few Q'eqchi' women leaned to use the new and unusual machines, several needles broke. Then a foot lever on one of the instruments stopped working. "The machines are asking us for something," declared one Qeq'chi' woman. Another reasoned that *costumbre*—a complex assortment of traditional Maya Catholic "customs" or ceremonies—hadn't been performed for the sewing machines, and so problems were naturally bound to happen. Soon many of the village women and their husbands met to discuss the matter, and concluded that proper *costumbre* rituals indeed were necessary to avert further trouble. The rites dictated that the sewing machines and the spirits inhabiting them be placated and "fed." With their rhythmic movement and ability to manufacture, the sewing machines were seen as obvious living entities, possessed by dissatisfied spirits known in Q'eqchi' as *xkwüinkul*, literally "its person."[*]

[*] The machines had been brought to San Lucas earlier that year by anthropologist

In the Q'eqchi' language, the ceremony is called *kwatesink*, literally "to feed." It is performed in many communities of the region in order to "socialize" what we would perceive as inanimate objects. Often, especially in modern times, the ritual takes place in association with newly made things, or instruments and objects brought into the community from afar. Newly built houses, tools, brooms, and musical instruments all may need to be placated through a *kwatesink* ceremony.[2] In San Lucas, the rite had also been performed on new ceramic stoves and for water tanks—important "living" objects introduced from outside the community.

So the ceremony for the sewing machines began one night, first with a visit to the church of San Lucas, where the women, their husbands, and other family members burned candles and incense. These initial steps took on an explicit cosmological significance. As the anthropologist Didier Boremanse notes:

> On one occasion the incense burner was placed on the ground in front of the church (to avoid having too much smoke inside the church) and four candles were placed in the four cardinal directions. All the people present were given a candle. A woman set fire to the incense and to the candles, and each participant lit his candle from his neighbor's. Everyone kneeled down facing the church and prayed aloud. The participants prayed toward the east. They stood up, and prayed facing successively the other cardinal directions.[3]

Later the group congregated at the house of the sewing machine's owner and gathered near the family altar decorated with images of saints. The incense burner was placed on the floor before the altar, one candle was placed atop the sewing machine, and three others on the altar itself. A procession of three or more women would then approach the

---

Didier Boremanse, after several women approached him with a formal petition requesting the machines, so that they could repair old clothes and make new garments. As Boremanse, 2000, writes in his fascinating account of the incident, he felt obliged to comply, and several machines were brought to San Lucas over the next few years.

sewing machine holding candles in their left hands. Each held in their right hands different offerings, including a burning *incensario,* a cup of cacao drink, and plates of food: tamales, tortillas, meat, and coagulated blood. The food was placed on the altar and even atop the sewing machine, and would later be eaten by the participants once the ritual had ended. The woman who led the ritual then "censed" the machine, ensuring that smoke enveloped the entire object. Although not explicitly mentioned in relation to this Q'eqchi' ritual Boremanse describes, it's no doubt significant that copal incense, called *pom,* is widely described by the Maya and other Mesoamericans as a sacred and fragrant "food" consumed by the gods. The act of censing the sewing machine can then be interpreted as one of direct "feeding." The women then continued to offer incense outside of the house at four different times, again to each of the four world directions, accompanied by Catholic prayers thanking God for the gift of the machine and its use. A male participant in the ceremony, a ritual specialist in the community, later repeated the process, censing the sewing machine, chanting the outdoor prayers, and making offerings to the cardinal directions.

The San Lucas ritual exemplifies a common pattern of Maya ritual, both modern and ancient, wherein inanimate objects (as we perceive them) are addressed and engaged as if they were living entities. The sewing machine, possessed by its troublesome *xkwünkul* spirit, required sustenance from the community in order to be calmed, placated, and socialized. The machines were outsiders brought into the village from a very different *ladino* world, and they therefore had to be brought into the community as social and moral *participants.* In this way, the souls of people and the souls of things could interact in peace with one another, extending social relationships beyond that of the people of the community.

The Eight Thousand Gods

For me, the San Lucas ceremony offers a compelling example of a very Mesoamerican notion that "things" often can assume a living essence, a true animistic quality. *Animism* is a word coined over a century ago by the Victorian anthropologist E. B. Tylor, who defined it very simply and far too generally as a "belief in souls."[4] Today, the sense of animism is more subtle and refined, suggesting a belief in ubiquitous souls attributable to non-human and even what we might call inanimate things. Seeing the world and its elements as alive or having an animate quality is a human phenomenon widely held by traditional and nontraditional societies the world over. The Shinto religion of traditional Japan is a highly animistic belief system, for example, that shows many remarkable parallels to some basic patterns in the Mesoamerican worldview and ways of thinking.

Carlos Lenkersdorf, an ethnographer who for years has worked and lived with the Tojolab'al Maya of Chiapas, Mexico, describes their animistic world simply and effectively. For the Tojolab'al, "everything is alive: people and animals, plants and springs, clouds and caves, light and wind, hills and valleys, stones and rivers, pots and griddles, crosses and roads, underworld beings and other supernaturals."[5] Along a very similar vein, anthropologist Alan Sandstrom, describing the system of beliefs among the present-day Nahua of Veracruz (descendants of the Aztecs), notes that "every hill, valley, spring, lake, stream, section of river, boulder, plain, grove, gorge, and cave has its proper name and associated spirit."[6] For much of traditional Mesoamerica, this animate quality of the world, where various important things possess a life force or soul, goes far to define things that are classed as especially sacred or godly.

It shouldn't come as much of a surprise, therefore, that in the period of early European contact, during the sixteenth and seventeenth centuries, notable features of the landscape were still places for veneration, sacrifice, and offering on certain days of the calendar. As one early Spanish chronicler described such "idolatrous" behavior among the Nahua in the early 1600s:

> Here . . . the Indians have the hills or springs, rivers, fountains, or lakes where they put their offerings on appointed days, like that of Saint John, that of Saint Michael, and other similar ones,

with the faith and belief that from those waters, fountain or hills
their good happenings and their health have their origin—or their
sicknesses if by chance those waters, fountains, or hills, or the
oliuhqui are angry with them, although it would be without their
having given them occasion.[7]

This focus on the living landscape resonates with very old notions of
urbanism and the design of communities, for in ancient Mesoamerica,
cities were landscapes in their own right, built up with artificial mountains, terraces, channels, and pools. In this way the urban landscape
was a conscious and deliberate reproduction of natural terrain and its
many "nodes" of sacred quality and expression. When processions and
prayers took place on temple shrines at Tikal, Teotihuacan, or Tenochtitlan, the architecture served as a reflection of this ubiquitous focus on
hills, springs, and other features of nature. The Mesoamerican notion of
the sacred encompasses many varied essences, spirits, gods and ancestors, many of which are at times difficult to categorize or distinguish
from one another. Even when we consider what looks to be a single,
well defined "character" in this system—an individual god or spirit, for
example—we quickly become aware that Mesoamerican cosmology and
religion could accommodate overlapping and merging entities and categories of being. The reason for this comes back to the inherent bound-
lessness of animism in Mesoamerican thought, where all manifestations
of the living sacred are interconnected, if not aspects of a larger whole.
As Eva Hunt has noted, "in [the Mesoamerican] view, as in those of all
pantheistic cultures, reality, nature, and experience were nothing but
multiple manifestations of a single unity of being."[8]

For the traditional Lacandon Maya who reside in the Chiapas rain
forest—they were until only recently un-Christianized—the ideas surrounding the manifestation of a single god named Aj K'ulel well illustrate these sorts of concepts. In their view, Aj K'ulel represents:

a single reality, like a gemstone whose different facets may reflect
red, blue, white, green or yellow light, according to the angle

from which one looks at it. Differences in meaning are but the reflection of different perspectives, and the perspective does not alter, or even touch, much less subdivide, the phenomenon.[9]

While it's true that the Mesoamerican worldview, including Maya cosmology, was based on such dynamic and malleable notions of the sacred, it's important to understand that manifestations of spirits and divine essences were not necessarily always fleeting, never to be re-experienced. Specific gods and spirits of nature might easily recur within this framework, and often in predictable and meaningful ways. In other words, there is logic and a system to the seeming chaos of multifaceted gods and beings, and it's this very ordering of such beings that defines various Mesoamerican cosmologies and religions.

Let's take, for example, the Maya deity known as Chaak, the god of storm and lightning. This all-important essence of rain is today

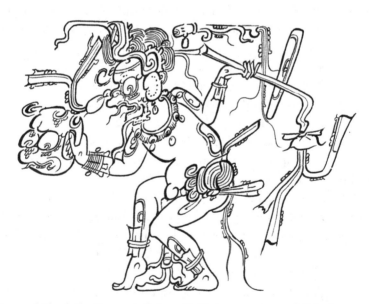

Image of Chaak the Maya rain deity, with his hammer stone and lightning axe.
(Drawing by the author)

still venerated in parts of Yucatán, and was called to the altar during a so-called Ch'a Chaak ("Get Rain") ceremony I witnessed and participated in as a boy (see chapter 4). Although often depicted in ancient Maya art as an individual god—a human with the face of a long-snouted serpent wielding an axe—it would be a mistake to think of him as simply one god who resides in the sky. Chaak was an individual with multiple aspects or personas, sometimes divisible into many at once. As a god of storms, he was associated with the four winds, and therefore had four directional manifestations: the Red Chaak of the east, the White Chaak of the north, the Black Chaak of the west, and the Yellow Chaak of the south. Of these, the most prominent was the Red Chaak, associated with the sunrise and the easterly winds that carried afternoon storms and, in some years, violent hurricanes. Chaak also had other specific aspects associated with different times of the year. Yax Ha'al Chaak was the god of the first rains, said in one ancient text to have been born at the start of the rainy season. Other "kinds" of Chaak included what was probably the spirit of misty rain, Muyal Chaak, and also the essence of heavy storms that brought out the first inundations of the year, Yax Bu'l Chaak, "First Flood Chaak." There may have been myriad Chaaks, each associated with some perceived phenomenon of nature or of the agricultural cycle. The storm god wasn't alone in having such varied forms and representations. The ancient Maya even seemed to have had a special term for this sacred collective: *jun pik k'uh*, "the Eight Thousand Gods," or, less literally, "the Multitude of Gods."

The words *teotl* in Nahuatl (the language of the Aztecs) and *k'uh* (*ch'uh*) in Mayan refer to such manifestations of the divine totality in the larger world. These terms, and their equivalents in other Mesoamerican languages, are nearly always translated simply as "god," but this is not quite the whole story. *Teotl* more correctly refers to a sacred and impersonal force or concentration of power, rather similar to the well-known concept of *mana* in Polynesia.[10] The Maya understanding of *k'uh*, or *ch'uh*, seems to be similarly wide-ranging, for it existed in many types of objects and people that took on a sacred quality of some sort. In some communities to this day, the word *ch'uh* forms the basis of an important concept of "soul" that could be possessed by all things important and valuable to the community. This life-giving soul of things and people is called *ch'ulel*, based on the word for "god" or "holy." Examples of what can possess *ch'ulel* include domesticated animals and plants, salt, houses and household fires, crosses, the saints, musical instruments, maize, and all the other deities in the pantheon. My late friend and colleague Evon Vogt, who with his students spent decades studying Zinacantan religion and society, once insightfully remarked that for the Tzotzil Maya, the most important interaction in the universe is not between persons and objects, but between the innate *ch'ulel* souls of persons and material objects.[11] Social dynamics are defined through the very idea of a shared sacred essence that pervades the earth, the sky, and the things and beings that inhabit it.

In ancient Mesoamerica, blood served as the material substance or vehicle for this sacred soul or life force. The Aztecs referred to the blood of human sacrifice as *teoatl*, "holy water," probably in the sense that sacrificial blood was the essential "drink" offered to sacred beings. Blood sacrifice is often thought to be the hallmark of Mesoamerican ritual and religion, and indeed it was important. For the Aztecs it was an extension of militaristic ideology involving sacred warfare, powerfully integrated into ideas about cosmology and world origins. In their myth of how the world began, it was the sacrifice of the gods that gave birth to the current sun and its associated age. In a self-fulfilling cycle, sacrifice begets creation and creation begets sacrifice. The Maya performed human sacrifice as well, although not quite on the same scale as the Mexica-Aztecs, with their centralized imperial state. Often captured Maya warriors and nobles were displayed in public and ritually executed as offerings to the gods. In all of Mesoamerican history, human blood served as a means of channeling and infusing the world with the sacred essence or "soul."

Here we enter into Mesoamerican metaphysics, and it goes far beyond concepts of animism and sacred essences. States of being were able to change, shift, and manifest themselves in remarkably complex ways. In a particularly insightful study of concepts held by the Tzutu-jil Maya of highland Guatemala, anthropologists Robert Carlsen and Martin Prechtel have singled out a key operating principle that drives universal processes of change and transformation.[12] In the traditional religion of the town of Santiago Atitlan, on the shores of Lake Atitlan, this concept is called *jaloj-k'exoj*. The two words *jaloj* and *k'exoj* in this pairing might both be translated as "change," but they have very different senses. As Carlsen and Prechtel describe it, *jal* is "the change manifested in the transition to life through birth, through youth and old age, and finally back into death." In essence this refers to visual differences or changes in one's outward persona. *K'ex*, by contrast, has the more specific meaning of "substitute, exchange." Among the Tzutu-jil, it refers to a complementary idea of change over the generations, as recurring people are "exchanged" for one another through repetition and reincarnation. It's through the process of *k'ex*, for example, that a Kiche' Maya child may assume the names of a long-deceased grandparent.[13] As Carlsen and Prechtel insightfully describe it, the paired notions of *jaloj-k'exoj* "form a concentric system of change within change, a single system of transformation and renewal."[14] The same system they describe appears in other Maya communities, and I suspect it has a very old and widespread history among the Maya. Among the Ch'ol Maya, the cognate terms *jel* and *k'ex* mean, respectively, "distinct, different" (that is, perceived differentiation) and "change, replace" (as an action, over time).[15]

These notions of cosmic change and recurrence are keys to how we approach and begin to understand many Maya and Mesoamerican rituals. In ceremonial settings, these are more often expressed as processes of "renewal," but the idea is much the same. While it may at first seem self-contradictory, many "rituals of renewal," be they agricultural rites or household ceremonies, are repetitious and "unchanging" ways of instilling ordered change and transformation. In Zinacantan, for example, saint images in household shrines must be ritually changed every fifteen days, as must the dying flowers that accompany them. Similarly, the installation of a new cargo holder begins with the redecoration of a household shrine in a ceremony called "Changing of the Flowers in the House."* The mayordomos and their assistants open their prayer before two altars, stating

> *The changing of the flowers is finished*
> *The changing of the leaves on your tree is done*
> *Now we have arrived at your great feast day*
> *Now we have arrived at your great festival*
> *You will be entertained*
> *You will be happy*

The new flowers and leaves on the cross (tree) are visual beautifications and surface changes (akin to the Tzutujil idea of *jaloj*) that revivify the cross in preparation for the arrival of the new cargo holder, the "substitute" in the recurring official role (as in *k'exoj*).

If the world operates according to these subtle ideas of change, it stands to reason that history, too, might have been conceived in much the same way. Past events were thought to exhibit patterns of cyclical renewal, replacement, and substitution, following many of the same rules found in nature and in ritual. That is, the concept of *jaloj-k'exoj* that Carlsen and Prechtel describe can be applied to the fundamental structure of the ancient calendar, where both linear and cyclical times are intermeshed as a system of "change within change." *Jal* seems to correspond well with the linear aspects of time as it changes in one direction, passing through the day names of the 260-day calendar, for example (see chapter 5). *K'ex* implies that specific names that appear in the linear time will recur, manifesting themselves again after intervals both brief and long term. Not only is *jaloj-k'exoj* a key mechanism for understanding the relations of humans within the natural and social worlds, but it also seems to operate as a fundamental structural and mechanistic concept in the Maya worldview, manifesting in numerous

---

* Vogt, 1969. This book on the multiyear study of social change and ritual in Zinacantan, Mexico, is a classic work of Mesoamerican ethnography.

ways. As we'll see in our later discussion of ancient Maya Creation mythology, these two terms play an important role in our understanding of how the ancient Maya saw the transformations of their world from the mythic to the historical.

The Solar Arrangement

For the traditional Ch'ortí Maya of southeastern Guatemala, the world is reborn every February 8. That date is the beginning of their New Year, and one of the most important days in the ceremonial life of various communities. Around this day "the Lord," the winter sun, long "resting" and dormant in its southernmost extension along the horizon, begins its gradual movement northward, bringing with it warmth and the promise of new crops and sustenance. February 8 is also, for the Ch'ortí, the formal start of the 260-day cycle, which survived well into the twentieth century in some parts of Guatemala. Over the decades and centuries the Ch'ortí have come to understand the 260-day round not as a continuous, recurring cycle, but instead as a fixed subdivision of the solar year, its start always anchored on this important winter date. These 260 days represent the period of most intensive work in the Ch'ortí agricultural fields, commencing when the sun begins its "walk" to the north.[*]

The New Year ceremonies begin with morning prayers and a ritual

---

[*] The Ch'ortí Maya today number about twenty thousand people, most living within a small area to the east of the modern city of Chiquimula, in the arid mountains near the Guatemalan border of Honduras, not far from the famous ruins of Copan. (For an excellent recent study of the modern Ch'ortí, see Metz, et al., 2009.) The Ch'ortí language is closely related to that found in the hieroglyphic writing of the Classic-era Maya (see Houston, Robertson, and Stuart, 2000). They may well be direct descendants of far larger populations that inhabited the lowland in ancient times. My description of the Ch'ortí ceremonies for the New Year are taken from classic accounts by the ethnographer Rafael Girard, who lived among the Ch'ortí in the 1950s. Many of the rituals he saw and described are no longer performed. See Girard, 1966, 1995.

breakfast for nine community elders, who convene in a village temple or shrine. There they make confessions before a Saint Francis, invoking the Virgin Mary as well as "the Child," the deified personification of young maize. Afterward, five of the nine elders participate in a pilgrimage to the west, walking in file at a brisk pace through the hills and arriving after some hours at a sacred pool. This journey of the five elders replicates the westward journey of the sun on this first day; the pool is an opening in the earth, symbolically equivalent to the entrance through which the sun passes into the netherworld. The principal elder ("The One in Front") is the true representative and personification of

A Ch'orti' Maya priest making offerings at a new-year ritual, photographed in the 1940s by the Guatemalan ethnographer Rafael Girard.

the sun, and of this first day of the calendar. The four other elders who follow the principal leader represent four gods of the cardinal points; they are "(the) four powerful men, the four miraculous men, from the four directions, from the four corners of the world, in the nominated spot (the pool), in the greater gathering place of the four Powerful Ones, of the four Miraculous Ones, of the four Angels." A woman who accompanies the five men follows at the end, embodying the moon as a conceptual counterpart to the One in Front.

Beside the sacred spring, the men and the woman take out a small tablecloth and place it on a cleared space upon the ground. On this surface, said to represent the plane of the cosmos, the men evenly place five straw rings that will serve as supports for vessels containing a sacred chocolate-maize drink known as *chilate*. These are arranged in a square-shaped diagram of the cosmos, with four rings in the corners and one in the center. According to the elders, each vessel is a symbol of a distinct "sun." The four points and the center are not oriented to the cardinal points, but rather to the four solstices—two on the east and two on the west, with the central point representing the intersection of the sun's path on the solstices. The point at the upper right of the quincunx corresponds to the sun's emergence on summer solstices, and the point at the upper left, its setting into the earth on that same day. The two points at lower right and left similarly represent the same emergence and entrance points of the winter solstice, closer of course to the southern horizon. At the center is the point of the earth, the cosmic "heart" that is the vantage point of all who perceive the motion of the sun across the horizon.

The Ch'ortí men place offerings around this makeshift altar—bundles of tamales, bread, and cheese—and two of the elders sit and face one another, praying and calling to the cosmic gods to come down to share their "obligation." The five cosmic entities then descend to earth and become embodied by the elders, who, through eating, enact a sharing of their divine offerings. The request is for the gods to provide ample rain and good health for the community, to fulfill their aide of a perpetual bargain between the people and the supernatural.

The Four Paths

Found throughout Mesoamerican art and geometric design, the form of the quincunx serves to highlight the idea of centrality that is very important in Mesoamerican thought, spatially defining a focal point of the sacred and a place of human perception. But rather than being a "simple map" of the cosmos, the quincunx is an elemental diagram of operational principles that are visible throughout the world, on many scales. Significantly, the bodies of humans and animals possess the same four-part symmetry as a quincunx. The two arms and two legs can be likened to the four outer points, centrally focused on the heart at the center. This link between the body and the geometry of the cosmos is a fundamental aspect of Mesoamerican religious belief.

This two-dimensional form, if visualized horizontally, is an ideal matrix used to understand and arrange conceptions of universal space and time, and it manifests itself in a variety of ways within the daily and ritual life of Maya communities. A house with its four corner posts, a rectangular cornfield, or *milpa,* or a community plan with four entrances, is each an expression of the quincunx from everyday life. The human or animal body, with its four limbs and a heart, is another

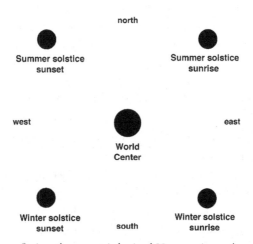

The quincunx design, the geometric basis of Mesoamerican solar cosmology.

that proves important in Mesoamerican thought.[16] By using and reproducing this form in a variety of ways, these and other spaces were microcosms in their own right, or small-scale replications of the cosmic template. While the four points of the quincunx and the "sides" that are formed between them are related to the concept of the four cardinal directions, they are also the numerology of Maya time.

The origins of the quincunx symbol can be traced back to Olmec art, to a glyphlike motif that represents the earth, often with a maize plant growing from its top. These geometric forms are found on many votive axes made from jade or greenstone, carefully fashioned by the Olmec and buried as offerings in large geometric patterns underground. The shiny stones with these designs of earth and cosmos perhaps symbolize sacred maize cobs or maize seeds, planted to ensure a regeneration of life and sacred essence within temples and courtyards.

Because of its symmetry and simplicity, the quincunx is in essence an intuitive shape in human cognition, and it finds expression in cosmologies well beyond Mesoamerica. In the American Southwest, cosmology was likewise represented in a four-part directional arrangement. In traditional China as well, the ideal of the center permeated people's worldview, political symbolism, and community planning in a way that ancient Mesoamericans would probably have appreciated. The center of the quincunx was, in the words of the noted Sinologist Paul Wheatley, a "pivot of the four quarters."* Its enduring symmetrical arrangement is still found today in Maya descriptions of their cosmos and in their thatched-roof homes and their maize fields with four corners and a center, always marked as sacred places in house dedication and cornfield rituals, respectively.

Many years ago I participated in a Maya rain ceremony at the village of Coba (described in detail in the next chapter), where my family lived for a number of months, and I can vividly recall how the rectangular form of the altar made by the village priest incorporated

---

* See Wheatley, 1971, for a classic look at the intersection of cosmology and community in ancient China. Many of the general patterns he describes resonate strongly in Mesoamerican thinking as well.

exactly the same cosmic symmetry. Bowls of sacred *balché* and *saka'* drinks were placed in the center of the table and at its four corners.[17] It wasn't until much later, however, that I was able to understand the more subtle ways that the humble-looking makeshift altar served as a microcosm reflecting solar movement and structure. Years later I read with fascination the published studies of my colleague William Hanks, who has for many years worked among the Maya of Yucatán, studying the ways in which their language expresses fundamental notions of space, time, and social relations.[18] He, too, participated in rain ceremonies, and was able to describe in great detail the symbolism of temporary altars nearly identical to the one I saw at Coba. The altars are usually rectangular in shape, but they are oriented in two different ways according to two distinct phases of the ritual. During the first phase, the altar is placed along the path of the sun, but with the longer dimension of the altar being north–south, thereby marking the movement of the sun between the solstices. In the second phase, at which the Chaaks are "lowered" to consume offerings of sacred bread, there is a central altar and two side altars, all with the corners oriented to south, east, north, west; the *hmeen* shaman performs from the northwest side and faces the southeast side. In this phase, arches of vines are constructed over the altars. The main altar has the vines rising from the corners and crossing over the middle; the lateral altars each have arches of vines over three sides, with the southeastern side left open, providing easy access for the arriving gods to consume the bread. Hanks suggests that these altars replicate the manner in which the Maya laid out their maize fields, making the layout and logic of altars iconic embodiments of agricultural space.

These quincunx and related four-part schemes of cosmology reference the same fundamental idea of "four-ness" important in Mesoamerica and well beyond. The famous anthropologist Alfred Kroeber once remarked, in fact, that four was *the* American number, meaning that it was found to be a key element in cosmologies throughout native North America.[19] There's no better description of the basic importance of four than that written by the Spanish priest Hernando

Ruiz de Alarcón, writing in 1629 about the role of the number among the Nahua people he came to know:

> Their use of the number four in all of their superstitions and idolatrous rites, such as the insufflations that they make when the sorcerers and false doctors conjure or evoke the Devil, the cause of which I was never able to track down until when I heard the story of the waiting for the sun. And for the same reason, the hunters, when they set their snares in order to catch deer, give four shouts toward the four parts of the world, asking for favor, and they put four crossed cords over a rock. The bowmen call four times to the deer, repeating four times the word *tahui*, which no one today understands, and they cry four times like a puma. They put a lighted candle on the tomb for a dead person on four successive days, and others throw a pitcher of water on it for him on four successive days. And, finally, among them the number four is venerated.[20]

Perhaps it is not at all surprising to know that in the Maya visual system of recording numbers using the portraits of various gods—even numbers were animate—"four" is always shown as the face of the sun god, K'inich Ajaw. The same association existed among the Aztecs, who considered their sun god, Tonatiuh, to be the patron of the same number. This obviously must derive from the sun's natural association with the four solstice points on the horizon.*

In Classic-era Maya art and writing, the quincunx is a hieroglyphic sign that stands for the word *bih*, meaning "road" or "path." This might seem an odd association, but I suspect that the connection derives from the Maya recognition that the quincunx—by their time already an ancient design—depicts the primary solar movements, the "four paths" that the sun takes on the seasonal solstices. We find confirmation of this

---

* See Gubler, 1997, for an overview of the importance of four in the Mayan worldview. Coggins, 1980, analyzes some of the iconographic aspects of four-part designs and conceptions in Mayan art.

The sun god with the glyph "four are his paths" (chan u bih k'in) atop his head, from a stela at Caracol, Belize. (Drawing by the author)

idea in a representation of the sun god on a sculpture at the site of Caracol, Belize. Atop his head, where we often find a name glyph, is a hieroglyph: *chan u bih*, "four are his paths." It may well be, too, that the idealized design of Mesoamerican towns and cities, consisting of four entrances and roads, is meant to replicate this basic cosmic idea of the sun's movement.[21]

There's some debate among Mayanists about whether we're actually looking at *directional* concepts in these four-part schemes, comparable to our own notion of horizontally fixed cardinal directions at east, west, north, and south. The evidence would suggest that this is not actually what the Maya and other Mesoamericans had in mind when they represented four cosmic points, whether with a quincunx or in some other format. As some have proposed, it could also accommodate a different understanding of the sun's movement across the sky over the course of a single day, resulting in four different kinds of "points": the place of sunrise on the horizon; the highest point of the sun during the day, the sun around noon, known as the zenith; the place of sunset on the horizon; and lastly a corresponding "nadir," or a point where one assumes the sun reaches its lowest point in its movement in the Underworld, before it begins coursing upward toward sunrise. To many scholars, these are the real "four points" of the Maya cosmos, not the cardinal directions as Westerners understand them.

The debate over this went on in the pages of academic journals for a number of years. I recall reading each article and each counterargument with great interest, wondering if I needed to do some serious rethinking of my own fundamental views of Mesoamerican cosmology. But I found it interesting that I agreed strongly with arguments on both sides. As it turned out, at least in my own thinking, the debate among scholars about directionality, zeniths and nadirs, and so forth stems from the ability, even the desire, of ancient Mesoamericans to represent four cosmic points in different ways, emphasizing *different sorts* of movements and categories of space to suit their purposes. This included, I firmly believe, something similar to our notions of four cardinal points, much as we see shown on the page of the Codex Fejervary-Meyer (see page 130), an ancient calendrical handbook from central Mexico that survived the Conquest.

So, we have the quincunx, with its representation of the solsticial points and the center, and a notion that that Mesoamericans were keen to emphasize sunrise, zenith, sunset, and nadir. These may seem like two very different notions of spatial arrangement, but in fact they are both valid and meaningful. One emphasizes the daily movement of the sun around the earth; the other is yearly movement of the sun along the earth's horizon. As the ethnographer Barbara Tedlock has noted, "Maya directions are *not* discrete cardinal or inter-cardinal compass points frozen in space but rather are sides, lines, vectors, or trajectories that are inseparable from the passage of time."[22] Tedlock's description is an apt one, but I would argue that Mesoamerican cosmology was equally suited to represent four sectors of the cosmos that do correspond roughly to our ideas of cardinal points. With main horizon points at the east and west, it was natural, I think, to impose a four-part scheme with regard to the horizon, with corresponding areas we call "north" and "south." Modern Mesoamerican peoples don't necessarily include terminology emphasizing the sun and its movement in their everyday

 east

 north

 west

 south

Glyphs for the four world quarters. (Drawing by the author)

speech, but ancient priests and artisans were surely well aware of the arcane cosmological concepts revolving around the sun, and were often happy to emphasize them in their writings and representations of the world.

Some of the first Maya hieroglyphs ever deciphered were those for the four directions, corresponding to east, north, west, and south. In a brilliant epigraphic advance, the French scholar Léon de Rosny noticed how, in a late Maya manuscript known as the Madrid Codex, a set of four hieroglyphs were often grouped together or were laid out in a diagram suggesting four cosmic points. Two of these glyphs, always opposite each other, included the *k'in* sign, meaning "sun." This gave de Rosny a vital clue, for the words in Yukatek Mayan for "east" and "west" are *lak'in* and *chik'in*. He was also able to tell which was which based on color associations he could discern in some of the surrounding hieroglyphs. So, for example, one of the two *k'in* glyphs was consistently associated with another sign that he knew was the color "red" (*chak*) and the other was found with "black" (*ek'*). Given the color associations of the cardinal points, he could easily say which glyph was *lak'in*, "east," and which was *chik'in*, "west." These were among the most important early insights into Maya decipherment, and they set the stage for the phonetic breakthroughs that would take place decades later.

In Mesoamerica and among many other cultures, as just mentioned, each of the four cardinal directions or quarters was associated with a particular color. Remarkably, this general connection of colors with world directions or quarters is widespread in the Americas and East Asia, leading some to think such cosmological ideas are extremely old in Native American thought, perhaps reaching back many thousands of years.* While there's some variation in the actual color schemes used by different Mesoamerican peoples, the system is surprisingly consistent across cultures of the area. For the ancient Maya, we know the following basic relationships between direction and color. (Here the Mayan words correspond to Classic-era terminology.)

---

* For a comparative analysis of color-direction associations among North American Indians, see DeBoer, 2005.

east: *elk'in* = red (*chak*)
west: *ochk'in* = black (*ik'*)
north: *xaman* = white (*sak*)
south: *nohol* = yellow (*k'an*)

Perhaps to this can be added the notion of the center of the quincunx as a fifth "direction," but it is difficult to establish that the spatial center was conceptually equated with the others. At least two of the four colors, red and black, have natural and obvious associations with their corresponding directions. The redness of sunrise may well account for the connection of red with east, for example, just as black holds an obvious link with the setting sun and the onset of night. Beyond this, however, it is interesting to see that red, black, white, and yellow are the colors of the principal varieties of maize cobs.[23]

A fascinating historical account of directional symbolism comes from the pen of Diego García de Palacio, who, as auditor of the kingdom of Guatemala in 1576, wrote a report to King Philip II of Spain on the conditions in southern Guatemala, then including the region of what is now Honduras and El Salvador. In that work he included the following extended description of a pagan sacrificial ceremony, notable for its narrative detail:

Besides their cacique or secular lord they had a kind of pope, called *tecti*, who dressed in a long blue robe, and wore on his head a diadem, or sometimes a mitre embroidered with many colors, at the crown of which rose a cluster of very beautiful feathers, taken from a bird called in this country *quetzal*. This pontiff carried in his hand a staff which resembled the crook of a bishop, and he was obeyed in all spiritual things. After him, next in sacerdotal authority, was the *tehu a matlini* who was the ablest diviner and the man best versed in their ancient books and in their arts. He it was who made auguries and foretold future events. After him were four priests called *teupixquis*, who went dressed in long robes, falling to their feet, each of a different color, black, red, green and yellow. These were the councilors of the pontiff, and directed all the superstitions and follies of their religion. There was also a kind of *mayordomo*, who had charge of the sacred jewels and the instruments of sacrifice. He also opened the breasts of the victims of sacrifice, and tore out their hearts, and performed such other personal services as were requisite. Besides all these there were other functionaries who played on the drums, trumpets and other instruments used in convoking the people to the sacrifices. . . .

Each year they had two principal and very solemn sacrifices; one at the commencement of the summer, and the other at the beginning of winter. These were made in the interior of the sacred place or temple, and were of boys between the ages of six and twelve years, bastards, born among themselves. . . .

They sound their trumpets and drums for one day and night before the sacrifice, and when the people were assembled, the four priests came out from the temple, with four small braziers in which they burnt copal and *caoutchouc;* and the four together, turning in the direction of the rising sun, bent their knees to it, offering incense and reciting words of invocation. After this they separated and did the same in the direction of the four cardinal points, south, east, north, and west, preaching and explaining their rites and ceremonies. When the sermon was finished they retired within four houses or chapels which were built at the four corners of the temple, and there rested for a little while. They next went to the house of the high priest, which was close to the temple, and took thence the boy who was to be sacrificed, and conducted him four times around the court of the temple, dancing and singing. When the ceremony was finished the high priest came out of the house, with the second priest and mayordomo, and ascended the steps of the temple, accompanied by the cacique and principal Indians, who, however, stopped at the door of the sanctuary. The four priests next seized the victim by his extremities, and the mayordomo coming out, with little bells on his wrists and ankles, stabbed and opened the left breast of the boy, tore out his heart, and handed it to the high priest, who put it in a little embroidered purse, which he closed. The priests received the blood of the victim in four *jícaras* (gourds), which are vessels made from the shell of a certain kind of fruit, and descending after one another into the court, sprinkled it, with their right hands, in the direction of the cardinal points. If any blood remained over, they returned it to the high priest, who put it back, with the purse containing the heart, into the body of the victim, which was interred in the temple itself. This was the kind of sacrifice made at the opening of the two seasons of the year.[24]

One wonders if the "two seasons" of the year are the dry and wet seasons, or perhaps the two solstices. At any rate, this lengthy description of a sacrificial rite at the time of Spanish contact (one heavily influenced by central Mexican terms and symbols, it seems) vividly shows how these spatial concepts played out in ancient ritual experience. The basic symbols of this ritual—four lines of young priests, the four points of the world directions, sacred vessels set in the world quarters—are not at all distinct from the Ch'ortí ceremony at the sacred pool, or from the Ch'a Chaak rain ceremonies of Yucatán I've witnessed. They are not the same ceremony, obviously, but they all reference the same sacred structure—what I think can be called the armature of the sacred—in order to happen the way they do.

The Layers of the Cosmos

Ancient Mesoamerican cultures shared numerous fundamental ideas about the universe, its forces and structure, and the relationship of humans to this larger divine world. Time is but one dimension in the complex and interrelated philosophy of the universe shared by the Maya, Zapotecs, Aztecs, and other cultures, and core ideas of this philosophy are still strongly held in traditional communities of southern Mexico and Guatemala.

Generally speaking (for there is some variation from area to area), the Mesoamerican cosmos encompassed three basic realms, all interrelated, in a total conception of the universe. The most important and directly relevant in human experience was the earth's surface, where humans, animals, and plants coexist and maintain a reciprocal and balanced relationship. One present-day Nahua man described to anthropologist Alan Sandstrom some of these basic ideas of the earth in somewhat forceful language:

> The earth is alive. We spend our entire lives defiling the earth, pissing on her and that is why she gets annoyed. That's why she wants to forsake us. Some people truly don't know any better . . . All things of value come out of the earth, even money. And yet here we are disturbing the earth, occupying it and planting on it all through our lives. Well, the earth can get annoyed because we disturb it. We plant beans, corn, sugarcane, bananas and camotes. Whatever it is we plant it on the earth. We go back and forth to market on it, and we get drunk on it but we don't give the earth any beer. We don't give her sugarcane, we don't give her bread, we don't give her coffee, and we don't give her joy.[25]

For the lowland Maya of Yucatán and environs, the centerline of the world, or axis mundi, takes the form of a world tree, ordinarily a ceiba, or kapok, tree, whose thick, tall trunk grows through the center of all the layers of the upper, middle, and lower world. To this day many small towns throughout Yucatán display a large sacred ceiba in the center of town plazas, or perhaps beside a local cenote, or sinkhole, which provides essential household water. In the Guatemalan and Chiapas highlands, the centerline of the world is often located in a ceremonial center, at a sacred shrine. This can also be a world tree, often in the form of a tall pole or cross located in front of the town church. In the Tzotzil Maya community of Zinacantan, this centerline is called the *mishik' balamil,* "the navel of the world," and is a small mound with a sacred shrine in the ceremonial center. In the Tzutujil Maya community of Santiago Atitlan, the axis mundi is called *kotsej juyu ruchiliew,* "Flowering Mountain Earth." It is a place at the world's center whose primary manifestation is a maize plant or tree. My colleague Robert

Carlsen, who spent much of his career living and studying in Santiago Atitlan, reports that before there was a world

> a solitary deified tree was at the center of all there was. As the world's creation approached, this deity became pregnant with potential life; its branches grew one of all things in the form of fruit. Not only did gross physical objects like rocks, maize, and deer hang from the branches, but so did such elements as types of lightning, and even individual elements of time. Eventually this abundance became too much for the tree to support, and the fruit fell. Smashing open, the fruit scattered their seeds, and soon there were numerous seedlings at the foot of the old tree. The great tree provided shelter for the young "plants," nurturing them, until finally it was crowded out by the new. Since then, this tree has existed as a stump at the center of the world. This stump is what remains of the original "Father-Mother" (*Ti Tie Ti Tixel*), the source and endpoint of life. The focus of Atiteco religion is in one way or another oriented backward, to the Father-Mother, the original tree. This tree, if properly maintained, renews, and regenerates the world . . . as long as the primal ancestral element, the Flowering Mountain Earth, is "fed" it will continue to provide sustenance . . . this "feeding" can be literal. For example, some Atitecos will have an actual hole on their land through which offerings are given to the ancestor . . . this hole is called *r'muxux* ("umbilicus"). More commonly, "feeding" is accomplished through ritual . . . dancing sacred bundles, burning copal incense, or praying can feed the ancestral form.[26]

In the Catholic church of Santiago Atitlan, the beautiful wooden altarpiece is carved in the form of the Flowering Mountain Earth capped by a maize plant. A few yards in front of the altar is a hole in the floor, usually capped, that is referred to as the "umbilicus" of the world.[27] Such remarkable fusions of Christian and traditional Meso-

american symbolism remain commonplace in parts of rural Mexico
and Guatemala.

Another layer of the cosmos below the earth is the Underworld, called
Mictlan (Beside the Dead) by the Aztecs, and *Xibalbá* (Place of Fright)
by the Maya. This exists deep in the earth's interior, a fearful place of
destruction and of uncontrolled, malevolent forces. Nightmarish spooks
and bringers of disease also dwelt there, ready to be conjured up onto
the earth through the forces of witchcraft. These Underworld entities
were not too far removed from everyday human life, where illness and
death were commonplace, if not routine. Such forces could also be ac-
cessed through direct experience by making pilgrimages to deep caves,
where prayers were made to the earth lords and other spirits. Many
such caves remain sacred in Mesoamerican religion to this day as sites
of pilgrimage and veneration of ancestors and earth spirits.*

The earth's surface, in some mythical accounts from the Aztec and
the Maya, was likened to the back of a great reptile—either a crocodile
or a turtle—bridging the realms of water and sky. The Aztecs referred
to this reptile as Cipactli, and the Maya of Yucatán, as Itzam Kab
Ayin. The earth was also perceived sometimes as a more anthropomor-
phic figure, an "earth lord" or an "earth goddess" (sometimes both),
especially among the cultures of highland Mexico such as the Aztecs,
where it was called Tlaltecuhtli, "Earth Lord." It is his face that likely
gazes from the center of the famous Aztec "Calendar Stone" or "Sun
Stone." The ancient Maya didn't have an exact counterpart to this idea
of a living earth being in human form. In their art they instead emphasized the animate quality of the substance of the earth—its stones
and mountains—by representing the face of living rock, called *tuun*, or
the spirit of mountains, called *witz*, shown usually as a large grotesque
mask with a prominent snout and large eyes, and often with maize
billowing from its top. Throughout much of Mesoamerica, mountains
were often seen as living beings that contained within them the spirits
of dead ancestors, deep beneath them in the Underworld.

---

* See Knab, 1995, for an excellent firsthand account of the power of witchcraft in
present-day Mesoamerican social and political life.

The animate face of a mountain, witz. (Drawing by the author)

The sky is the third component in the Mesoamerican triumvirate of cosmic space. The daytime sky, resplendent and blue, was considered to shine like the fine jade or turquoise jewels of a king. The Maya called it *chan* or *kaan,* and often depicted it as a plane or band of abstract celestial symbols. Sometimes the arc of heaven could be depicted as if it were a large serpent's body, probably likened in some manner to the path of the daytime sun. The words for "sky" and "snake" (*chan* and *kan*) are near homophones in Mayan languages. In the art and iconography of the Classic-era Maya we sometimes find the sky represented as a frame containing an array of jade jewelry, probably meant to convey brightness and vitality. It's reminiscent of ideas still found today, for example, among the Nahua of Veracruz, who see the sky (*ilhuitl*) as an immense shiny mirror "filled with the brilliance and sparkle of the sun and stars."[28] Another cosmic crocodile served as a symbol of the night sky, complementing the great reptile that formed the earth's surface. This was most likely a representation of the arc of the Milky Way.

According to most Mesoamerican cosmologies, the sky was composed of thirteen levels or zones, where specific deities resided. There's no doubt this is related to the importance of the number thirteen in the structure of the main Mesoamerican calendars, the Calendar Round, and in the

Maya Long Count, as we will see in later chapters. Space and time were conceptually united in a way Einstein would have appreciated.

In Mesoamerican perceptions of the cosmos and space, the world also sometimes assumed the metaphorical form of an architectural space, a "house" constructed and sustained by the gods.[29] Numerous native descriptions of cosmic forms and structures emphasize this literal architectonic notion of the sky as a roof supported at its four corners or sides. As my friend Evon Vogt once wrote to me, "All things, people, animals, and features of the landscape occupy and conduct their affairs in this cosmic structure." The house metaphor might only seem a superficial spatial parallel, but it is possible that this correspondence also involved a broader metaphor or conceptualization that likened the world and its social dynamics to a "household"—a cosmos that was defined not just by its formal similarities to building, but also by a sense of community and social identity. In fact, it is sometimes difficult to differentiate the physical and social notions surrounding the Maya "house," even on a real-world scale. We know, for example, that the "house" was of great importance as a social and even territorial unit in ancient Mesoamerican thought, and the interweaving concepts it encompasses included the idea of a town or community.[30]

When speaking of a cosmic architecture, therefore, I don't refer only to the visual concept of the world and how it is put together, but also to the fundamental ways in which the world encompassed ideals of social and even political relationships on a wide range of scales. Put very simply, the Maya and other Mesoamericans have long perceived a porous "flow" among the metaphorical concepts of the living body, the house, the community, and the world as a whole. And as we've already begun to discern, firmly integrated into this network of cross-referenced concepts and animated structures is time itself. The ancient calendar was also a living, multilayered construct, an expression of space as much as of the passing of days. It's the stunning depth of that temporal dimension to the world that we will now explore.

4

fINDING ORDER

The first grand discovery was time, the landscape of experience. Only by marking off months, weeks and years, days and hours, minutes and seconds, would mankind be liberated from the cyclical monotony of nature. . . . Communities of time would bring the first communities of knowledge, ways to share discovery, a common frontier on the unknown.

—*Daniel Boorstin*, The Discoverers

No rain fell early in the summer when I was nine years old, living with my family in a remote Maya village in the eastern part of the Yucatán Peninsula. A dry, parched jungle is an odd setting to experience, even for a kid who has never before seen or heard the rain forest. I still remember it all so vividly, almost four decades later. The birds were silent, the bugs were few, and the sun was relentless. Fear and tension had spread among the people of Coba, the village where we lived, for corn planting time was fast approaching. Without rain, normal life for the people of Coba—and for us—would be impossible.

Those months during which we all lived in a thatched house among many Maya families was my first exposure to the rural life of the contemporary Maya. For two years my father oversaw the mapping and survey of the extensive ruins in the forest surrounding Coba for the National Geographic Society. The place is now visited by tourists, who stay at Cancun and the other beach resorts along the overdeveloped "Riviera Maya." Back then, in the 1970s, the sprawl of Cancun was still years away, and Coba could be reached only by a rough dirt track throughout the forest. Everyone in town spoke Yukatek, a close relation to the royal court language of the Itzá at Nohpeten. We were, for a short time at least, part of the community's social fabric, though no doubt a bit oddball in the neighbors' eyes. We kept dogs and chickens as pets (and not to eat or hunt with), drove around in a big Chevy Suburban, and played Beatles music from a tape deck. But our family fit in in our own way, and the experience forever shaped my life, beginning my lifelong love for the Maya people and their history and culture.

Due to the drought that summer of 1975—our second year there—I saw just how vital the cycles of the two tropical seasons, wet and dry, were to village life. In a season that normally saw daily cloudbursts each and every afternoon, the storms never came. As my mother later remembered, "clouds built up, turned dark; and, helpless, we saw them drift out to the north and out to sea." The previous year, we had survived Hurricane Carmen as it passed almost directly overhead our pole-and-thatch hut—which, I might add, makes for a reasonable shelter in a hurricane, since the rain and wind go through the walls; you

The author at nine years old resting among the ancient ruins at Coba in the summer of 1975. (Photograph by G. Stuart)

are wet, but intact!—so the contrast from the first year to the second was memorable and jarring to my nine-year-old mind. So bad were the dry conditions that by the end of June, the men of town, unable to plant their corn seeds, sought desperate measures by holding an elaborate rain ceremony to pray to the ancient storm deities. The Maya call this ritual Ch'a Chaak, roughly "Get Rain," Chaak being the name of the animate spirit of storms and lightning. The ceremony was to last through the night and most of the day, and we Stuarts were invited to participate, including, to our great surprise, my older sister and my mother; women are typically forbidden from participating in the Ch'a Chaak for fear of their polluting sacred ground. Our close connection to the community and to the people of Coba had by that time allowed for a change in the customary rules.

The previous year, my father had hired one of the young men of Coba to assist with his archaeological mapping. He would use his machete to cut paths in the jungle as sight lines for the surveys of the thousands of ruined structures that lay scattered in the rain forest. His name was Jacinto May. I remember as a kid hearing that Jacinto was training to be the *hmeen,* or shaman, of the Coba village, and then wondering what exactly a shaman was. I must have asked my father or mother about it, for I somehow came to learn that Jacinto was in training to learn the esoteric knowledge of Maya gods and the prayers for curing sickness and communicating with things supernatural. As a kid who was at first bored with the simplicity of life in Coba—I later came to love it—this sure sounded interesting. As I later learned, a *hmeen* is literally a "doer" or "maker," and often one of the most important individuals in a traditional Maya community. Since Jacinto was only a trainee at that point, an elder, more experienced *hmeen* was brought from a neighboring town, Chemax, to oversee the communal ceremony.

The day before the evening portion of the ceremony was to begin, a handful of men spent the morning building a temporary altar in a clearing surrounded by the fallen ruins of Coba's massive temples. I vividly remember the eerie and magical setting. I watched them as they worked and quietly conversed in Mayan, curious and a bit apprehensive about what the ensuing ritual for the rain god would be like. The men used rustic wooden poles to make a four-sided ceremonial table, with thin tree branches arcing from corner to corner on each side. Pit ovens were dug nearby, into the ancient stucco floor of the temple, in order to cook the ceremonial breads that were to be offered to Chaak and then eaten by all the participants. A number of boys from Coba watched along with me, many my age, perhaps also curious about the ceremony that most had probably never seen. As I was to learn in a few hours, some had already been selected to be performers in the ritual.

All of us were required to bring offerings to the Chaaks. We asked Jacinto what we should bring. "Coca-Cola would be good," he told my father, "also perhaps some cigarettes, if you have them." My mother went to the local *tienda*, a few huts away from our own, to buy enough of these items for the ceremony. Coke and cigarettes might seem strange offerings for a celestial spirit, but it's important to remember that, back then, these were much-prized imported commodities, exotic in their own way and therefore fitting for making requests to the god of rain.

We gathered that night around the finished altar, and soon the prayers began. We all sat quietly as the older *hmeen* started his chants, taking sips from a gourd bowl containing *saka'*, a sweetened corn drink. Our faces were lighted by a fire built off to the side, and by the flames of the pit ovens. The droning of the *hmeen*'s prayers was utterly incomprehensible to me, and soon seemed endless as I began to tire. The *hmeen* and Jacinto would pause, and a new round would begin soon thereafter. During one of the prayers, late into the night with the moon overhead, four boys, some my Maya friends, knelt at the sides of the square altar and made the sounds of small frogs: *woh, woh, woh, rana, rana, rana*. Their surreal and mesmerizing chants, my father told me, were calls to four Chaaks, living at the four corners of the world.

More drinks were shared as the night wore on, but no longer did we partake of the sweet corn drink; now our beverages were concoctions known as *balché*, an awful-tasting fermented liquid made from the bark of a tree and sweetened with honey. Our hosts, knowing we gringos preferred our Coke, decided to add the carbonated and sugary drink into the *balché* to make it more palatable. I quickly learned that carbonated *balché* isn't much of an improvement on standard *balché*—in fact, I can recall its unpleasant bittersweet, fizzy taste even now.

And as we all learned soon enough, this *balché* recipe was also a strong purgative. My brother, Gregg, my sister, Ann, and I were terribly sick to our stomachs for the entire week that followed. Luckily we had access to the only modern bathroom in town, graced with plumbing, not far from our house, in order for our bodies to rid themselves of their impurities. Sometimes I found that short walk was nearly impossible in my weakened state; a few days after the ceremony, my mother found me on the trail to the restrooms, unable to stand, with my head wedged in the V-shaped branches of a small tree, just my height. But after a few more days we recovered.

And about a week later, I think it was, the Chaaks came, and it rained.* Looking back, I now see how those summers at Coba helped me to understand the concept of time as something real and experienced, untethered from our more "normal," abstract sense of minutes, hours, weeks, and months. Simply living in a world without electricity allowed for the routine passage of alternating days and nights to deeply affect my sensibilities in new and unexpected ways, even for a child. In much the same way, the constant outdoor life we led at Coba allowed me to see remarkable changes in the longer term. With the coming of that first rainstorm in June 1975, the environment transformed overnight. The din of thousands upon thousands of singing frogs was deafening that first night after the rains fell, and the insects, of course, were also resurrected from their deep sleep. In the lowlands of the tropics, the transformation of the seasons is a remarkable thing, so unlike the more gradual changes I experienced back in the United States. It's

---

* My mother wrote and published her own remembrance of the Ch'a Chaak ceremony, which of course filled in numerous gaps in my *balché*-influenced memory. This can be found in *The Mysterious Maya* by G. Stuart and G. Stuart, 1977. For a discussion of similar Ch'a Chaak ceremonies elsewhere in Yucatán, I highly recommend Love, 2004, and Hanks, 1990, for detailed descriptions of the ceremony's symbolism. Kintz, 1990, has written a brief and compelling ethnography about the town of Coba, which is now fully integrated in the growing tourist economy of the region.

one of the things I still love about living in that part of the world, where there's no question that these sudden comings and goings of nature's cycles played a role in shaping the Mesoamerican conceptions of time.

Another important insight from that experience, now that I look back on it, was seeing how people who live outside of our conventional hour-by-hour rat race nevertheless strive to exert some control over the forces of time and seasons. The Maya of Coba, living in those days on the cusp of age-old tradition and modernity, were desperate to restart a cyclical system that was seriously out of whack, and communication with those animate forces of time and nature was therefore of paramount importance. These desires to impose order out of chaos undergird the way all cultures structure, arrange, and represent time, sometimes in quite different ways.

For most people, "time" is not just an experience of nature and its built-in cycles of days and seasons. It's more of a human construct, a way of ordering and structuring those natural cycles to make them conform to our needs and daily experiences. This shepherding of the days and seasons leads to much more elaborate and necessary constructs we call calendars. Our calendars may appear at first to be "natural"—who says there aren't twelve months in the year?—but history tells us that even our own system of organizing time was forged over many long centuries, and involved many complex and even random reasons. Such is the case with most other calendars, even the ones devised by ancient Mesoamericans, as they attempted to find meaning and order in their experience of time's passing.

Time's Elements

Forget for the moment that there are seven days in a week, that there are twenty-nine, thirty, or thirty-one days in a month. Forget that there are twelve months in a year, or ten years in a decade, or ten decades in a century. Most of us today take these numbers and subdivisions of time as absolute givens, rigid and unshakable as we pass through them balancing checkbooks, meeting deadlines, and making resolutions to do better next time. But if we delve deep into the origins of these numbers by which we live our lives, we see that some turn out to be surprisingly arbitrary, while others, such as the lengths of a day, a month, and a year, are astronomically dictated and experienced directly through our senses and perceptions. For the sake of discussion, then, I suggest that we discard much of the temporal framework upon which we schedule and organize our lives and experiences; that framework needs to be put aside so that we can truly appreciate how varied human thinking about time has been over the course of the four to five millennia we have available for study.

It's easy to assume that all our own notions of time and its passing are simply a reflection of the dry mechanisms of astronomy, a reckoning of days and months as the earth and the moon predictably dance about each other in their constant orbit around the sun. To some extent this is true, but our sense of time is also very much influenced and conditioned by our biological selves, resulting in a very human perception of these mechanisms that is seldom ever perfect or exact. We all know enough about time to make use of clocks, calendars, and datebooks, yet those structures we call days, weeks, months, years, decades, and so on are in large part constructs of culture and history, fashioned upon the knowable realities of the sky and the laws of physics. Different cultures at different times have tried to impose certain imperfect yet often elegant structures on those dispassionate mechanics of the sky and seasons, and our calendar is one of many created over history, in different places on the earth.

We tend to think of calendars as things—an array of numbered squares on an office wall, a schedule book in one's pocket, or, more increasingly these days, a computer program by which we organize our activities and responsibilities. But it's important that we look past such representations of time and strive to examine calendars and day reckoning in a more abstract way, as *conceptual* structures of time. That is to say, it helps to understand that calendar systems are not necessarily the same from culture to culture. Each system for tracking the passage of time, including our calendar, always exists as part of a certain mind-set. The way in which a culture organizes time is always inseparable from its wider context, from powerful ideas about cosmology, history, and experience. In this way, cultures the world over have developed myriad calendars throughout prehistory and history, if only a small percentage of them have survived to be analyzed and examined by modern anthropologists and historians of science.

Far from being things, then—that flat, rectangular object hanging on an office wall, say—calendars are ways of organizing our experience relative to the perceived mechanisms of the world and the cosmos as a whole. Timekeeping didn't develop simply as an offshoot of early human attempts to tally things or to write, that is, as a recording method. Rather, it is representative of a shared mental conception of how time "works," and in this respect, calendars are found throughout the world's cultures, even among illiterate societies once considered primitive.

A few basic facts about astronomy and celestial mechanics are key for understanding how we humans "make time," in the literal sense. Our varied calendars and structures of time find common ground in the sky, as it were, by acknowledging a few universals about the movement—or perceived movement—of prominent heavenly bodies such as the sun and the moon. All calendars, including our own, arose in the depths of history as combinations of observed astronomy and somewhat arbitrary tallies of days, months, and years with roots in the variables of history, religion, and human thought.

Even in our own daily experience, the way we engage with time is a mirror of both science and culture. Take the day, for example. We conceive of it as the twenty-four-hour period it takes for the earth to make one complete pivot on its polar axis. On most of the earth's surface, this involves a period of direct exposure to the sun (daylight) in combination with a dark period (night) while the earth's surface is turned away from the sun. In English, *day* usually refers to the period of light, which was the word's original meaning—*day* is a cousin to the Sanskrit word *dah*, "to burn"—but this association was later extended to mean an entire twenty-four-hour span.

The day is obviously basic to everything. It sets biological rhythms throughout our world, among countless species of plants and animals. These are known as circadian rhythms, and we see them at work everywhere around us, even if we mostly take little notice of them. Life on earth (most of it, anyway) relies on the energy and warmth of the sun to process nutrients and drive growth, so daylight becomes the driving factor for existence itself. Sunflowers bloom and face toward the sun as it travels across the sky. It might seem so obvious as not even worth mentioning, but we humans are likewise slaves to our biology as it's been shaped by the movement of our planet—we have to sleep, and we have evolved to do so at night, which, for us, makes daytime almost synonymous with experience itself.

Let's begin with a clock, that most basic and pervasive of human inventions. The twelve hours that we see on the face of a clock have no basis in real astronomy. Rather, they probably have very distant origins in the Mesopotamian and the classical worlds, corresponding to the twelve divisions of the zodiac and their associated constellations, through which the sun was perceived to move throughout the day, from sunrise to sunset, over the course of a lunar month. Before the Late Middle Ages, only the hours of daylight were subdivided in such a way—why keep track of night so rigidly, after all?—although the Romans had standardized the middle of the night as a convenient and workable transition point between the days themselves. It was only with the revolutionary (literally) invention of the clock and its precise mechanisms that the dark period of night succumbed to the same twelve subdivisions of the daytime, giving us the two twelve-hour cycles of any clock, and the day of twenty-four hours. Although astronomy and astrology were companion endeavors in the classical and medieval worlds, it's important to stress that the twelve-hour subdivision of our day is artificial—an artifact of culture.

The seven sequential days of our week likewise came about through a compromise between astronomy and the constant human desire to impose structure on time. As their labels show, the days were named originally after the seven main heavenly bodies. Sunday and Monday are clear enough, though other languages make use of Anglo-Saxon names for deified planets, overlaid on an older Roman system, which was itself derived from a widespread Mediterranean calendar traceable to the Babylonians of more ancient Mesopotamia. Once we make the parallels, our days are, or were, dedicated to (in order) the sun, the moon, Mars, Mercury, Jupiter, Venus, and Saturn. (Uranus and Neptune, fine planets both, came to be discovered far too late, only after the invention of the telescope, to make for a nine-day week.) This seven-part system may have been seen to dovetail roughly with the length of a full lunar cycle, which varies from twenty-eight to thirty days, so that four such "weeks" compose more or less a month. Again, we see that the period of seven days is not based on any astronomical pattern or reality; it is a cultural construct, an attempt to assign personalities to the days through use of the main celestial bodies known in the classical world.

As we'll see in the next chapter, Mesoamericans recognized many more than the seven days to which we're so accustomed. In their sacred day count there were separate names for twenty sequential days, many with labels based on animals or important elements or forces of nature. No close parallel with the names of the classical seven-day system would be expected, of course, but I find it interesting that both traditions, along with many others, chose to designate the alternating periods of light and dark that humans experience with terms borrowed from other aspects of nature and experience. As a basic time period, days are the least concrete and material of concepts, but humans across time and history have preferred to name them and lend them something approaching a character or personality. For the Maya, as we will see, this was especially important.

As human societies past and present considered higher periods of time, it was natural to lock the passing days in to the very visible cycles of the moon, or "months," and even the larger units we call years. It's reasonable to think that these two essential concepts of time—months and years—were among the first large-scale time units that humans discerned in a scientific way, as they observed the sky even hundreds of thousands of years ago. The moon's cycle is perhaps what we see recorded on the earliest man-made "calendar" that has come down to us, as markings etched into a bone by an early inhabitant of what is now southern France more than thirty thousand years ago. The meaning of this artifact is still debated, but there can be little doubt that the people of that very remote time were keenly perceptive of the moon and the sun, and of the way the world changed around them on a regular basis. Their calendar, like ours, was in many ways simply experiential.

The technical term for a lunar cycle of roughly 29.5 days is a *synodic month,* and nearly all calendars of the world make use of this basic unit. Each page of one of our wall calendars represents one of these natural subdivisions of time. The English word *month* is based on the word for "moon," as is the case in a great many languages of the world. In Mayan languages, for example, the word *uh* is "moon" and is also the period of roughly twenty-nine or thirty days. So, like us, the ancient Maya realized that the synodic month did not correspond to a whole number of days; there was some give in how they defined a month's duration. They made use of an elaborate lunar calendar in order to anchor their written historical narratives. These routinely accompanied records of k'atun endings or of important ritual dates, in the form of a standardized set of hieroglyphs known as "the Lunar Series," which specified the moon's age on a given day—say, the birth date of a king. Records reveal that the Classic-era Maya liked to record these prominent dates using a lunar calendar that specified the number of elapsed days from the moon's "arrival"—what we call the "new moon"—and that they placed synodic months into groups of six, so that there were roughly two full lunar cycles per solar year.

The lunar month of 29.5 days has an intimate connection to biological rhythms here on earth, although some aspects of this remain poorly studied scientifically. The average menstrual cycle in human females lasts 29.1 or 29.5 days, and cultures the world over have naturally come to see a direct connection between this aspect of human reproduction and the moon.* The word *menstruation* is based on the Latin word *menses,* for "month." Is this a coincidence, or did human biorhythms evolve

* See Cutler, et al., 1987, "Lunar Influences on the Reproductive Cycle in Women," *Human Biology* 59 (6); Walter Menaker and Abraham Manaker, 1973, "Lunar Periodicity in Human Reproduction: A Likely Unit of Biological Time," *American Journal of Obstetrics and Gynecology* 117.

in association with the relative movements of the earth to the sun and the moon? Scientific opinions differ on the matter, but it seems likely that a connection does exist. Many peoples have also understood, naturally enough, that nine lunar months corresponds with the period of human gestation. (In Mesoamerica, as we will see, this natural period from conception to birth may have formed an important basis for the sacred divinatory calendar lasting 260 days, or about 9 months.) It's probably no coincidence that the moon is often viewed as a female being in traditional mythologies and folktales from many regions, including Mesoamerica. Among the ancient Maya, for example, the animate moon was seen as a beautiful young woman, often a companion to the god of the sun or the sky.

Traditional farmers know that moon cycles can be essential for timing the planting and harvesting of crops. And numerous cultures know that the moon can even affect human moods and behavior (among lunatics, for example). So while the jury may still be out on questions about specific connections between the moon and life here on earth, there is no question that, like the sun, the moon can have a profound effect on biological cycles on our planet. These may be rooted in even more archaic cycles of reproduction among species whose biorhythms are intimately associated with the ebb and flow of the ocean tides.[1]

As with the names of the seven days of the week, the labels we use for our months derive from Latin names in the Roman calendar. The Romans based much of the reckoning of time on the unit of the lunar month, and over time they added new layers of complexity onto what started as a fairly simple system. Perhaps as early as 700 BC, centuries before the founding of the Roman republic, a festival calendar was established based on traditional agricultural cycles that no doubt had an even more distant and archaic origin. Some names are from familiar Roman gods (Janus, Mars, and Juno, for example), while others are from well-known emperors (Julius and Augustus). Originally there were 10 such lunar months, which spanned 304 days, beginning in March (Mars's month) and lasting until the end of December, soon after winter solstice. The remaining two winter months were of little concern to early Roman agriculturalists, who left their fields fallow until planting in the spring. Calendar reforms came often in the ensuing centuries, and eventually the two winter months, *Januarius mensis* and *Februarius mensis*, gained official recognition, making for twelve in all. With the reforms of the so-called Julian calendar, efforts were made to make the lunar months fit nicely with the length of the solar year. Here one runs up against a basic conundrum that has frustrated chronologists since the dawn of history: given that the synodic month of 29.5 days, or even a rounded month of 29 or 30 days, does not go evenly into a solar year of 365.24 days, some sort of fudging is necessary. The Julian calendar reforms of 45 BC led to the insertion of one extra day every fourth year—what we call the leap year—in order to make up the difference. This is precisely the calendar system we've inherited in the Western world.

The history of calendars shows a strong inclination to have the lunar months conform and intersect with the solar year, as shown by the simple manner in which numerous calendar systems emphasize the familiar "fit" of twelve moons within a 365-day year. Twelve months of 29 or 30 days get one almost there, to about 360, a "vague year" period that the Maya happened to find especially useful, even though they also had full knowledge of the more exact solar year of a bit more than 365 whole days. Our own calendar relies on the same basic idea, but does so by means of rather messy and inconsistent lengths for the months. As the familiar centuries-old saying goes:

> *Thirty days has November,*
> *April, June, and September:*
> *Of twenty eight is but one,*
> *And all the rest are thirty-one.*
> *Of course Leap year comes and stays,*
> *Every four years got it right,*
> *And twenty-eight is twenty-nine.*

I suspect an ancient Maya priest would have hated this hodgepodge manner of structuring the year's subdivisions—we'll see later how the Maya and other Mesoamericans preferred a different way of divvying up 365 days into "months" that in fact had little outwardly to do with lunar phases and cycles. But the twelve-month system we use was arrived at long ago, and came to be employed by cultures throughout the ancient world. Egyptian, Hindu, Islamic, and Old English calendars are all very similar to ours in structuring the year as twelve "moons," even if they came upon slightly different means of forcing the lunar and solar cycles to conform to one another. Such systems are a reflection more of the universal human need for structure than of natural reality. The length of the synodic lunar month as twenty-nine or thirty days derives from the length of time it takes for the moon to orbit the earth, plus a slight adjustment in time of about two days to account for the earth's movement relative to the sun. In terms of celestial mechanics and physics, however, this length of time from new moon to new moon has no direct correlation with the solar year, which is all about the earth's full orbit around the sun and its seasonal tilt relative to our home star. Human calendars have tried for eons to squeeze these two different but fundamental phenomena about the movement of heavenly bodies into a cohesive system, even where one does not exist.*

As we'll explore in considerable detail in the next chapter, Mesoamerican "months" based on the synodic period of twenty-nine or thirty days were not quite so emphasized in the everyday calendar, or at least not in the written forms of the calendar that have come down to us. For the Maya, the year was roughly divided up into two sets of six lunar months, although there seems to have been little if any effort to force an imperfect fit of this scheme into the solar year. In other words, the lunar months and the solar year, while each important to ancient Mesoamerican timekeeping, did not seem to be woven together into some

---

* The two lunar months and the solar year do converge more neatly when considered on a bigger time scale. The so-called metonic cycle of 6,940 days (about nineteen years) represents a common multiple of the solar year and the synodic month.

sort of artificially imposed system. Months were really not *part* of the solar year; they operated concurrently but largely in separate realms.

According to the modern scientific definition we all learn in school, the year of slightly more than 365 days corresponds to the duration taken by the earth to pass one full orbit around the sun. This movement of our planet within the larger solar system is the foundation of the Copernican model of the universe. But the idea of the year does not in any way rely on Copernicus, of course, and is instead based on age-old observations of the movement of the sun and of other celestial bodies in the sky. Humans have always been able to perceive the year in a very precise way, without knowledge of the mechanism of planetary movement in a solar-centric cosmos.

The year is, generally speaking, the largest *natural* time unit that we humans use in the organization of our lives and societies. Of course longer time cycles from astronomy are real and perceptible, but apart from a few learned astrologers or scientists, we humans have never really paid a whole lot of attention to them. Other long units of time that we routinely use—decades, centuries, and eras—are not categories or units of time that have any basis in astronomical mechanics; they are simply collection units of years that arose through our base-ten counting system, in order that historians and long-range planners could talk about larger and larger units of time. (In Mesoamerica, as we will see, all cultures made use of a base-*twenty* system of counting, including years.)

Years are basic, obviously, but when do they begin? Given the different ways human cultures define the starting and end points of a year, it's not as self-evident as it might at first seem. The ancient Romans traditionally began their year in the month of March, probably to roughly coincide with the spring equinox and its associated agricultural activities. Equinoxes, like the solstices, occur only twice a year, and happen when the sun is aligned directly above the equator. On these days, in March and September, the lengths of the day and the night are equal (*equinox* is from the Latin *equi,* for "equal," and *nox,* for "night"). Subsequent reforms brought the beginning of the year closer to the winter solstice of late December, which is seen by many cultures

of the Northern Hemisphere as a time of world rebirth and regeneration. In time, the first day of January gained status as the official beginning of the Roman year.

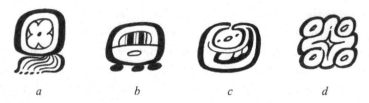

a   *b*   *c*   *d*

A selection of Maya time glyphs and terms (from left to right): K'in, *"day,*
sun"; Ha'b, *"year";* Uuh, *"moon"; and* Ek', *"star."* (Drawings by the author)

The Mayan word for "year" is *ha'b* or *haa'b*, which is probably based
on the word for "water" or "rain," *ha'*. Its literal origin may well have
been something like "wet season," which came to be extended to the
full 360- or 365-day period. (These spans we will explain a bit later;
suffice it for now to say that *ha'b* could refer to either.) In Maya glyphs,
the sign used to write the word *ha'b* is an oblong-looking element
shown above and labeled a which, depending on context, can stand
for a period of 360 or 365 years. The form probably originated as a
representation of a log slit-drum, a resonating musical instrument that
was of great importance in performance and ritual throughout ancient
Mesoamerica. The Aztecs called this type of drum *teponaztli*, and the
Maya, *tunkul*. It's curious that a drum would be a visual cue for the
word for "year." Perhaps it has to do with the likening of passing years
to rhythmic drumbeats. Alternatively, it may allude to the way that
Mesoamericans probably viewed the subdivisions of their year not as
simple "months" or other periods, but as different-named festival periods, all accompanied by much music and dance. I wonder if we can see
some hint here of musical rhythms providing a key metaphor for the
marking of time, as if its passage were somehow musical performance.
   Smaller heavenly bodies of the night sky—the pinpoints of light that
are, to us, planets and stars—can also play important roles in human
systems of time reckoning. The planets that are visible to the naked

eye, if carefully tracked over long periods, follow predictable movements in the sky and also in relation to the sun's rising and setting, for
example.

Of all of the planets, Venus is most intimately related to the sun, due
to its close proximity to it in the visible sky (and in the solar system, as
we know today). This brightest of all planets never assumes a very high
visible position in the night sky, because it can be seen only for a time
before sunrise, when the planet heliacally rises in advance of the dawn,
or after sunset, when it follows the sun into the western horizon. Due
to the quickness of Venus's orbit relative to that of other planets, the
timing of these phenomena can be easily tracked and predicted over a
relatively short period, without the need of a telescope. The Maya were
especially interested in Venus and its movement, and they included
schematic tables of its course in their sacred almanacs. In addition to
Venus, the only other planets one could see in the night sky were Mars,
Jupiter, and Saturn. There's little doubt that astronomer-priests among
the Maya and other Mesoamerican cultures kept some account of these
other planets, but the evidence of this is not too terribly solid.

So we have three natural ways in which human beings have long
perceived and structured time, based on routine changes in the sun,
the moon, and the seasons. These are the essential building blocks of
all calendar systems, whether our own or those from some remote period in history. Each represents something different: The day is how
we perceive the movement and cycle of the sun as it moves across the
sky, disappears, and reemerges again. The month is very different,
perceived as the amount of time it takes for the moon to pass through
one full cycle of transformation, whether defined as from full moon
to full moon, or from new moon to new moon. The year, in turn, is
perceived as the duration of a seasonal cycle, discernable by looking at
the sun as it moves along the horizon, reaching (in the Northern Hemisphere) its northernmost point on the summer solstice and its southernmost point on the winter solstice. Alternatively, like the ancient Greeks
and Romans, one could discern a yearly cycle by tracking the position
of the noontime sun relative to the southern horizon. Looking at the
changes over time in the length of a shadow cast by a perfectly upright

pole or stick is one useful way of perceiving the changes in the sun over the course of a year.

All of the calendars developed throughout human history, including that of the ancient Maya, saw days and months and years as separate elements in some integrated system, even if the schemes themselves varied a great deal over time and space. That is, we can agree that our notion of a "year" as a specific length of time associated with a regular, perceptible change in the seasons was probably not too terribly different from the concept of the year held by an ancient Roman, Egyptian, or Chinese, for instance. Humans throughout history would therefore probably agree on a common "vocabulary" of time, even if the calendar structures upon which these elements are built—the "grammars" of time—varied a great deal. In other words, calendar systems are not consistent, are even messy, and may bear little resemblance to one another from one culture to another, even if their internal components might be easily recognizable.

Forming and Re-forming Time

In the long sweep of Western history and science, there's been an incessant effort to improve time's measurement, and to perfect its usefulness in people's daily lives. Calendars evolve throughout history, sometimes fall by the wayside and are lost, or come to be invented out of whole cloth. This, if nothing else, shows that they are not inherently perfect or perpetual. Rather, they are always in some sense imperfect, sometimes even clumsy, creations of human minds and cultures. In addition, the various calendars that have been dreamed up over the course of human history show just how erratic these attempts at order can be. Calendars came and went during specific points of history, and were adapted for special political and religious purposes that may or may not be so relevant for other eras and cultures.

Our own means of timekeeping have gone through a number of significant and very intentional changes during their long history. The

Roman calendar, as we've already seen, provided the essential and familiar building blocks for the system we now use. In the early days of the republic, the year was of inconsistent length, some years being 355 days in length, alternating with other, longer "intercalary" years of 377 or 378 days. Careful adjustments were often necessary to keep this rather confusing system in line with the tropical or solar year of 365 days, though historical circumstances such as war sometimes got in the way of this operation. In 46 BC, therefore, Julius Caesar instituted a major calendar reform in consultation with his scientific advisor Sosigenes, as a matter of political and social policy. As we've seen, the lengths and numbers of months were adjusted slightly so that each single year was 365 days in length, and an extra day came to be added to every fourth year to create what we call a "leap year," in order to keep the year in agreement with the observed position of the sun. This makeover of the civic and religious calendar of the Roman state has come to be known as the Julian calendar, and it continued to be used for well over a thousand years of European history. The Romans, of course, did not reckon years relative to the birth of Christ—this began much later, in the sixth century AD, and gradually spread throughout Europe over subsequent centuries.

The second major reform to the Western calendar came centuries later, during the Renaissance. By the sixteenth century, a considerable number of errors and some imprecision had occurred in the proper placement of Easter in the Christian calendar—essentially a borrowed Julian calendar with Christian feast days included. Easter was especially tricky, since its timing was decreed in the fourth century to be inexorably tied to the moon—it is always the first Sunday after the fourteenth day of the lunar month that comes after the vernal equinox. After so much time, however, enough drift had accumulated in the relative timing of the week, the equinox, and the lunar month as to cause widespread confusion. Pope Gregory XIII instituted the change in 1582 by omitting ten days from the calendar: October 4 in the Julian system changed over the next day to October 15 in the Gregorian calendar. The new calendar didn't immediately take hold everywhere, however. Predictably, the Protestant authorities in northern Europe and in England were none too keen to adopt this "papist" time frame. It was therefore not formally instituted in England until 1752. The Gregorian reforms stand as the basis for the calendar we use today.

Some ambitious calendar reforms have turned out to be miserable failures. One of my favorites is the Republican calendar, or *Calendrier Républicain,* used, as its name suggests, for a short period in the wake of the French Revolution, when the new Revolutionary government opted to abolish the use of the Christian years in favor of a brand-new year count. In the Republican calendar, 1792 was declared "year I," 1793 "year II," and so on. Each year consisted of twelve months of exactly thirty days, each of which in turn consisted of three "weeks" of ten days each. The conventional names for the months were completely rejected in favor of more "natural" terms that reflected the agricultural cycles so important in the life of the common man: month names such as *Frimaire,* "Frost," and *Messidor,* "Harvest." At the end of the year, a shortened period of five days followed the thirty-day months in order to approximate the full solar year. Advocating a "purer" decimal system for organizing time, individual days were subdivided into ten-hour periods, each of one hundred "minutes," and each of those consisted of one hundred "seconds." New clocks were manufactured to follow this new and thoroughly different system. Amazingly, this calendar was formally adopted by the French government and was in place for a dozen years before its end was formally declared by Napoleon in 1806. Evidently this calendar reform, spurred by Revolutionary zeal for an entire new social order, ran up against considerable disinterest among the general population. It also proved a major problem for France's neighbors and trading partners, and its colonies in other parts of the world, where a month name such as Frost held little meaning.

It's fascinating to me that such a radical restructuring of time was ever even attempted. There was a keen recognition on the part of the calendar's designers, organized by the Revolutionary politician Charles Gilbert Romme, that the structure of time was something utterly shapeable and symbolic, even outright political. This has been shown many times over in the course of history, but perhaps no better than in those short-lived days of the Republican calendar. Of course, calendar reforms have existed well into recent times, and I suspect they will continue into the future, as politics and religion always strive to shape human experience. The controlling of time, be it symbolic or more grounded in reality, is a powerful symbol indeed.

Time in Mesoamerica

We've seen that the Mesoamerican calendars used by the Maya, the Mixtecs, the Aztecs, and others shared many fundamental features, but there were some internal differences among them. Some cultures, for example, opted to use or emphasize particular types of cycles over others at particular times. As we'll see in chapter 5, the Aztecs designated years by naming them after their first day in a 260-day divinatory calendar. Yet for the ancient Maya of Classic times, years were seldom ever shown this way, at least in official history. They instead preferred to make use of a place notation system that represented the accumulation of days and years from a set starting point. After the end of the Classic period, this "Long Count" system was dropped and time was represented in slightly different ways in the later written record. Such regional and historical differences in time reckoning within ancient Mesoamerica can be hard to sort out, but I suspect that some changes can be explained by conscious efforts to reform or reconsider time, driven in large part by societal changes in religion, ritual, and politics. This, as we've seen, isn't new or unique in human history. All cultures with long-standing conceptualizations of time regularly subject their calendars to change and refinement. In some ways, speaking and writing about calendars takes us away from the reality of those basic natural cycles that led to their creation in the first place. Today we seldom if ever think of our months in terms of lunations, or of the subdivision of our years as having any basis in agricultural seasons.

Calendars have always existed to bring order to individual lives and communities, and in historical perspective we see that the desire for structured time was always intimately tied to religion—another arena that desires to put order on what is otherwise a rather messy and incomprehensible world. Today we use our calendar to structure our work and education, and to plan our future projects and goals. We use it as a framework for various kinds of historical understanding, recent or remote. Only in the last few centuries has our time reckoning transformed so radically toward more and more levels of precision, as the need developed to navigate the seas with more accuracy, or to deliver manufactured goods "on time."

But in ancient times, in Mesoamerica and beyond, calendars were intimately tied to attempts to understand and divine nature as well as human fate. The need for temporal structure and organization remained a practical reality, and served as a necessity for organizing and making sense of the vagaries and uncertainties of everyday experience. Nowadays, as an archaeologist thinking back on my boyhood experiences as a participant in the Maya rain ceremony at Coba, I wonder what a ritual asking for rain might have looked like a thousand or two thousand years ago, when corn farmers in the same area had to deal with occasional droughts and communities were under similar dire circumstances. Ch'a Chaak rituals asking for rain surely existed then, too, in some form or another, in the remote towns and villages away from the great ancient cities. Maybe even the terrible-tasting *balché* was served to its participants.* In those remote times, unlike today, the shaman priests who oversaw the ceremonies were probably well versed in the messages and meanings of time as it was anciently structured—what Mesoamerican cultures called the "Order of Days"— using it as a framework to divine the reasons why the rain deities, and the rain itself, were so far away.

---

* *Balché* was no doubt an ancient ritual drink, being mentioned in early historical accounts as an important part of pre-Conquest Mayan ceremonies in Yucatán. See Landa, 1941, p. 92.

5

IDEAS OF THE DAY

What I shall deal with now is the count of the years, the days, the months, and weeks which governed these people in pagan times; the names and figures that designated the days to predict the fates, destinies, and inclinations of the newborn; the order of the calendar and the feast. . . . Regarding all of this there was careful computation, so that things had to take place on such and such a day or season and so that everything might be in its right time.

—*Fray Diego Durán*, The Ancient Calendar, *1579*

I was born four and a half decades ago on a day the Aztecs would have called *Nahui Ollin*, "4 Earthquake." This might sound bad, I'll grant you, but we'll soon learn that this combination of a number and a name has very important meanings and, for the Aztecs, even positive associations. My two sons, Peter and Richard, were born on days named *Nahui Ozomatli*, "4 Monkey," and on *Nahui Acatl*, "4 Reed," respectively. Our common connection to the number four is interesting—it's a numerical coincidence with only a 1-in-2.197 chance of happening—and I would like to think that a Mesoamerican day-keeper or diviner would have seen this as a meaningful connection among us Stuart menfolk. We've already seen that four is a good cosmic number in Mesoamerican thinking. Such odd-sounding designations for the days—numbers accompanied with *Earthquake, Monkey, Reed*, and so on—represent a basic and key component of all Mesoamerican time reckoning, a system so vitally important to daily life, in fact, that it remains in use to this day in certain areas of Mesoamerica, five centuries after the arrival of Europeans.

Each combination of a number and a name, such as "4 Monkey," represents a single station in a perpetual cycle of 260 days. The numbers span from 1 to 13, combining with one of 20 individual day names, producing a total of 260 possible combinations (13 x 20 = 260). The names themselves follow a set sequence, and simple mathematics requires that any individual pairing of number and name, such as 4 Earthquake, be repeated each 260 days. Before we delve into the specific day names used by the Maya and their neighbors, it may help to demonstrate this structure using letters to represent the twenty day names; let's call them days A through T. The numbers precede the names in most Mesoamerican systems, including the Maya system, so we can abstractly represent a sequence in the following way, again with each number and letter combination standing for a single day:

1A, 2B, 3C, 4D, 5E, 6F, 7G, 8H, 9I, 10J, 11K, 12L, 13M, 1N, 2O, 3P, 4Q, 5R, 6S, 7T, 8A, 9B, 10C, 11D, 12E, 13F, 1G, 2H, 3I, 4J, 5K, 6L, 7M, 8N, 9O, 10P, 11Q, 12R, 13S, 1T, 2A.

In this tally, the basic structure of the calendar is clear. The numbers and names each follow their predetermined sequence, yet the day-to-day combinations themselves change. I chose to begin this with "1A," but any starting point would involve the same idea of interlocking 1–13 and A–T in a perpetual sequence. It's an endless round.*

All Mesoamerican cultures that shared this 260-day cyclical calendar gave names to each of the twenty days, and, remarkably, many of their meanings show equivalencies or close correspondences across different Mesoamerican languages. For example, the seventeenth day name is *Ollin* (Earthquake) in Nahuatl, *Kaban* (Earth) for the Maya, and *Xoo* (Earthquake) among the Zapotecs. The same is true for many, though not all, of the other names. Thus the

---

* My father, George Stuart, first came up with the number-letter method for representing the basic structure of the 260-day round.

260-day round was similar across all of these cultures, pointing not only to close cultural connections, but also, presumably, to their common origin deep in Mesoamerica's past. As we'll see, the archaeological evidence of the 260-day round's earliest use remains patchy, making its first appearance around 400–300 BC, in hieroglyphic inscriptions left by the Zapotecs and the Maya. Its widespread use even then, well back into what we call the Preclassic period, leads me to think that the earlier Olmec and their contemporaries probably used this very same divinatory calendar, maybe as long ago as 1000 BC, if not before.[*]

The Aztec Days

The richest sources we have for learning about ancient Mesoamerican day reckoning come from early colonial period descriptions of the Aztec calendar, mostly in documents compiled by Spanish priest-scholars during the sixteenth century. We also have a number of original ancient books, known as codices, collected in various places in central Mexico during and after initial contact between Mesoamericans and Europeans. Most of these precious pictorial manuscripts, so brilliantly painted with colorful, intricate imagery, were priests' handbooks— almanacs full of religious symbolism and featuring a great deal of calendar-related content, including tables used for making divinations and prognostications. The basic elements of the Aztec calendar were very much the same as the system used by the earlier Maya, so a detailed knowledge of Aztec timekeeping offers insight into many features of Mesoamerican timekeeping now lost to us.

In the 260-day calendar of the Aztecs, the 20 days—what we designated simply by letters in the scheme just described—had the following names in the Nahuatl language:

---

[*] There are claims that the 260-day calendar was invented earlier than this (Malmström, 1997, p. 52), but no firm evidence exists to support such claims.

1. *Cipactli,* Alligator
2. *Ehecatl,* Wind
3. *Calli,* House
4. *Cuetzpalin,* Lizard
5. *Coatl,* Snake
6. *Miquiztli,* Death
7. *Mazatl,* Deer
8. *Tochtli,* Rabbit
9. *Atl,* Water
10. *Izcuintli,* Dog
11. *Ozomatli,* Monkey
12. *Malinalli,* Grass
13. *Acatl,* Reed
14. *Ocelotl,* Jaguar
15. *Cuauhtli,* Eagle
16. *Cozcacuauhtli,* Vulture
17. *Ollin,* Earthquake
18. *Tecpatl,* Flint-knife
19. *Quiahuitl,* Rain
20. *Xochitl,* Flower

As expressions of time, each individual name had little intrinsic meaning. Only with the addition of the 1–13 number prefix could they be used to designate a given day, or what was called a *tonalli.* I am writing these words on the day 10 Earthquake, so tomorrow will be 11 Flint-knife, the next day 12 Rain, then 13 Flower, 1 Alligator, 2 Wind, 3 House, and so on, in an endless interlocking progression of number and name. Purely by the math, as noted, the same combination of a number and a day name will not repeat for exactly 260 days.

Aztec glyph for the day "6 Rabbit."

As we see in the figure shown here, the hieroglyphs used by the Aztecs for these days are highly pictorial and mostly self-evident, typically accompanied by a string of dots that represent the number.

In the Aztec system, the day signs are all highly iconic: often the head of an animal or deity that corresponds directly with the day name. Attached to the sign, sometimes by a line, are strings of circles, never more than thirteen, each of course representing a "one"—unlike the Maya, the Aztecs did not employ bars for "five"—resulting in a line of dots that resemble beads of a necklace.

The Aztecs called their full 260-day cycle the *tonalpohualli*, usually translated as "the count of days." In essence, the full array of numbers with day names represented the essential working tool for diviners and soothsayers in Aztec communities, who studied their perceived meanings to foretell the future and to influence many aspects of daily life. The books they used were manuals of fate known as *tonalamatl*, "books of days," one of the best known surviving examples being the Codex Borbonicus. Although many often translate *tonalli* as "day," as here, the word more correctly refers to one's personal day sign, corresponding to the fateful day of one's birth in the 260-day system. In this way one's *tonalli* would have functioned not unlike our concept of an astrological birth sign, determining a great deal about one's fate, character, and role in society. One case illustrates the idea for us, described in detail in a native Aztec text compiled by the great sixteenth-century priest and scholar Fray Bernardino de Sahagún:

> [The day sign] One Flower. . . . The man born upon it, they said, and it was averred, would be happy, quite able, and much given to song and joy: a jester, an entertainer. And it was said that women were great embroiderers. It was said that this sign was indifferent; that is to say, a little bad and a little good.[1]

Upon birth or very soon thereafter, an Aztec child was ritually bathed and assigned his or her *tonalli* sign. Individual signs had either positive or negative associations (sometimes both), which led to some hand-wringing among parents and priests. For example, a child born on a favorable day would immediately be baptized and wedded to his or her life symbol. If the day was unfavorable—a "beastly day" or one "full of sin"—then there was some wiggle room in the scheduling of the ceremony. Aztec priests could delay the rite for a few days, as long as it fell within the extended "week" of thirteen days, called a *trecena*. So, for example, 1 Flower (*Xochitl*) begins such a period that ends on 13 Grass (*Malinalli*). If a child was born within this period on, say, 9 Rabbit (*Tochtli*), and if this was seen as undesirable, the baptism could be delayed until one of the four remaining days of the *trecena* period. Still, one can only imagine what took place when children were born near the end of a period, when few options for tweaking the newborn's *tonalli* sign would have been possible.* So intimate was the association of one's *tonalli* to the character of the individual that days served as given names for people, gods, and spirits, especially among the Aztecs and Mixtecs.

The *tonalli* was far more than a sign for one's birthday or a designation for the sun. The word also referred to the life force thought to reside in the head of every person, a force that was believed to give children the ability to grow and be conscious beings. The word literally means "the sun's heat," which by extension also signified the heat and life force of one's body. The *tonalli* (sometimes just *tonal* or *tona*) was thus one of the basic types of human souls conceived by the Aztecs, encompassing one's animate being, one's fortune or destiny, and one's essential character as an individual. Significantly, the word *tonalli* can be seen as part of the descriptive name for the Aztec sun god, Tonatiuh. The name of this god in Nahuatl, one of the richest and most expressive of languages, literally means "He who goes along becoming warm," referring to the course of the sun

---

* In the *trecena* system, a *tonalli* day such as 1 Flower or 1 Reed presided over the thirteen-day intervals that they inaugurated. A sequence of *trecena* stations would look something like this, where the number prefix goes unchanged (because of their thirteen-day intervals) but the day name changes:

1 Flower, 1 Reed, 1 Death, 1 Rain, 1 Grass, 1 Snake, 1 Flint-knife, 1 Monkey, etc.

These opening stations held important divinatory meanings that helped to determine something about the character of the entire *trecena* period that followed, until a new presiding day took over with the appearance of a new day sign carrying the number one as its prefix.

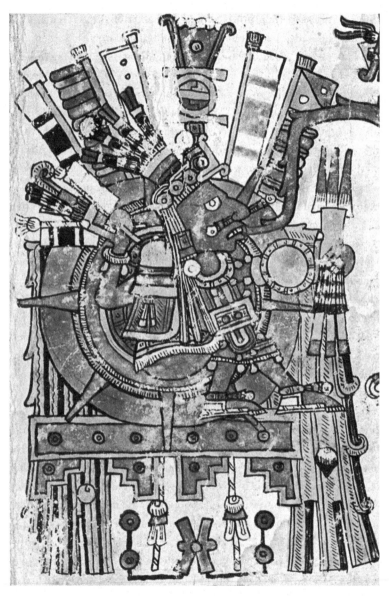

Tonatiuh, *the Aztec solar god, from the Borgia Codex.*

through the sky. Tonatiuh is shown in the art of the Aztecs and its neighboring cultures as a handsome young human, with golden hair and jade jewels, much like his Maya counterpart. My colleague Karl Taube has suggested that these highland cultures of central Mexico may have depicted him as a Maya nobleman, from the hot lowland country to the east, where Mesoamerican jade originated. Tonatiuh's brightly colored locks of hair reflected his brilliant light, and the gold metal (*teocuitlatl*) that decorated the temples and the bodies of the nobility was seen as his excrement. (Metal ore of all types was thought to come from the sky, like celestial droppings, an idea that may have originated from the observation of meteorites.) When Cortés marched through Mexico and into Tenochtitlan in 1519, the Aztecs took quick notice of his chief lieutenant, the notoriously cruel Pedro de Alvarado, and his shock of blond hair. They soon named him Tonatiuh, a warrior god, and the name forever stuck. In the Borgia Codex, a Nahua pictorial document perhaps collected by Cortés himself, we see Tonatiuh depicted as an enthroned warrior-king, backed by a radiant solar disc and drinking the streaming blood from a sacrificed bird. The giver of life and heat in this way consumes the hot life force of other beings, including, on many occasions, the blood of sacrificed people.

Looking closely at one page of the Codex Borbonicus—a prime example of a *tonalamatl* that was probably painted right after the Conquest—we can get a good sense of how priests would quickly have been able to access key information about the 260-day calendar. Many of the screen-fold books are composed of different types of almanacs and charts, providing visual cues about the days and their symbolic associations. Those devoted to the 260-day cycle are typically arranged into *trecenas*—that is, into 20 groupings of 13 day signs, each set represented on a single page. On one page of the Borbonicus, for example, we see an image panel depicting a seated goddess of water, Chalchiuhtlicue (Jade-her-skirt), with a gridlike array of squares to the bottom and the right. There are five columns along the page's bottom, each with two squares, and five rows along the right side, also of two squares. Together the columns and rows represent the thirteen days, with the

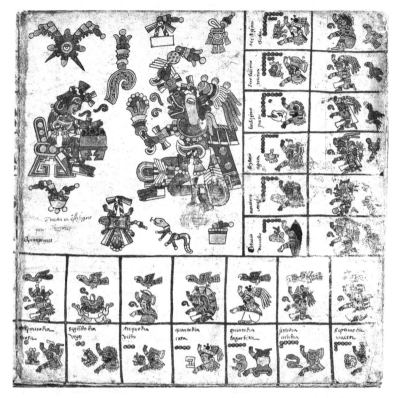

A page of the Codex Borbonicus, an Aztec day almanac, or tonalamatl.

sequence beginning at the lower left, running along the bottom, and continuing up to the upper right. Looking closely, we see strings of dots in one of the squares for each day, corresponding to the numbers one through thirteen. The numbers accompany different day signs (among other symbols) so that we have running along the bottom, left to right:

1 Reed (effaced), 2 Jaguar, 3 Eagle, 4 Vulture, 5 Earthquake, 7 Flint-knife, 8 Rain

and then up to the top, along the right:

8 Flower, 9 Alligator, 10 Wind, 11 House, 12 Lizard, 13 Snake

Accompanying the *tonalli* days in each of these small squares we find a deity with outstretched arms, as if protecting or embracing each day. There are nine different in all, again in a set sequence, so that the figure depicted with 1 Reed appears again with 10 Wind, after which the sequence of nine deities repeats. These are the so-called Nine Lords of the Night, who were said to have presided over the dark hours of each day, from sunset to sunrise. Each was an important god in wider Aztec religion, such as Tlaloc, the god of rain, or Centeotl, the god of maize. In the other small panels adjacent to the days we find other small figures of Aztec gods, thirteen different gods in all, accompanied by a different flying bird (or, in one case, a butterfly). These are the patron deities of the *trecena*, gods who were associated with one of the individual numbers. Tonatiuh is shown as the overseer of the number four on the day 4 Vulture, and Tlaloc, the rain god, again appears as the patron of the number eight, with the day 8 Flower. Unlike the *tonalli* days and the Nine Lords of the Night, these thirteen gods and their birds co-vary with the numbers, so they appear in set, unchanging positions in each of the twenty different Borbonicus tables.

A day priest used a page like this as a *tonalamatl* almanac for understanding the supernatural forces and influences associated with a given day. If he wanted to know what to prognosticate about a future occurrence of day 2 Ocelotl, for example, he would consult such a book and look up that day's prescribed associations: (1) its night lord is Tlaloc; (2) its number patron is Tlaltecuhtli, the Earth Lord; (3) its bird omen is a green hummingbird; and (4) it is part of the *trecena* sequence of the day 1 Reed, with all thirteen days presided over by the goddess Chalchiuhtlicue. We don't know the arcane details of what such associations mean, but it's clear the books gave priests the basic working tools for knowing and judging which influences were in play as they went about the procedures of proper soothsaying.

The power of prognostication went far beyond one's birth date. The scope and importance of the 260-day calendar is vividly conveyed in a passage from an early account of idolatrous beliefs among the Zapotecs, written in 1656—over a century *after* the Conquest—by the bishop of Durango and Oaxaca, Diego Díaz de Quintanilla y de Hevia y Valdés. From what we know from a variety of sources, his description probably would be fitting among the Aztecs and other Mesoamerican peoples as well:

Those they call learned men and teachers have long taught the same errors and false ways of their religion, for which they make use of books and manuscripts. They instruct others of thirteen gods that bear names of men and women, to whom they attribute various powers, such as in the system of their year, which is composed of 260 days, and that is in turn divided into thirteen months, to each of which is attributed one of the said gods who governs it. The year is also divided into four ages or thunderbolts, each of sixty-five days. With such sorcery the calendar priests derive a variety of magical spells and superstitions, regarding all types of hunting and fishing; the harvesting of maize, chile and grain; regarding all illnesses and the superstitious medicines used in curing. . . . They instruct also about when to stop working; how to keep the dead from visiting their houses; for success in pregnancy, in childbirth and procuring children; for the omens of bird songs and animals; for the interpretation of dreams and fixing the harms that they prognosticate; and finally for anything that occurs to any of these learned men and teachers.[2]

The skeptical bishop offers us a good sense of how few, if any, aspects of daily life fell outside the realm of soothsaying and interpretation anchored to the 260-day calendar.

Even more detail about the pervasive power of the calendar comes across in the writings of Fray Diego Durán, who spent nearly all his life living among the Aztecs near Mexico City in the decades following the Conquest: "The characters [for the days] also taught the Indian nations the days on which they were to sow, reap, till the land, cultivate corn, weed, harvest, store, shell the ears of corn, sow beans and flax-seed."[3] Durán went on to write that:

> these signs were common not only in agriculture but also in trade
> and commerce, in buying and selling, in marriage, and in bathing.
> The same was the case in the eating of certain foods; except on
> specified days and times certain foodstuffs could not be eaten. I
> believe this superstition to be difficult to uproot, and I fear that in
> certain places these ancient rules and rites have not disappeared. I
> see that they are still kept strictly, and I base my opinion upon the
> fact that one day I asked an old man why he was sowing a certain
> type of small bean so late in the year, considering that they are
> usually frostbitten at that time. He answered that everything has
> a count, a reason, and a special day.[4]

Durán was right to recognize the stubborn survival of the day count even in his time, the late 1500s, for, as we'll soon see, it still survives in some remote Mesoamerican areas up to the present day. In these sorts of early descriptions we also get a sense of how the Spanish took special notice of the power of time and the calendar over the daily lives of the native population. The 260-day calendar was far from some arcane religious calendar; it pervaded all aspects of daily life and shaped activities in ways that, as with the late planting of a bean field, would otherwise seem counterintuitive if not downright illogical. This often to the chagrin and frustration of the Spanish priests, who were determined to stamp out the power and influence of the native soothsayers. In one amusing episode, Durán recounted how one traditional day-keeper interfered in his church service:

> I dare to swear to these things because in church I myself have
> heard the public pronouncement, all the people being present,
> that the time of the harvest had come. They all rush off to the
> fields with such haste that neither young nor old remain behind.
> They could have gathered the crop earlier, at their leisure; but
> since the old sorcerer found in his book or almanac that the day

had come, he proclaimed it to the people, and they went off in great speed.[5]

Just who were these sorcerers and soothsayers? The Aztecs called them *tonalpouhque*, "*tonalli*-readers," and they were considered among the wisest and most important members of the community. As Durán says, the Aztec soothsayers he encountered in the sixteenth century were "wise in the old law, who taught and still teach the young folk who are now being educated . . . in the count of the days and the years and of the ceremonies and ancient rites."[6] We shouldn't necessarily

An Aztec soothsayer divining the fate of a newborn child; from Sahagún's Florentine Codex.

equate them with the elite priests or "god keepers" who oversaw activities within the temple precinct of Tenochtitlan. The *tonalpoubque*
were instead the widespread guardians of arcane knowledge, healers and diviners who could probably be found in communities both
large and small in the Aztec Empire. They were above all intermediaries between the world of the sacred and the everyday concerns of the
people, aiding in childbirth, curing illness, and divining the fortunes
of many other activities of daily life. And although Spanish writers
of the time often used the terms *soothsayer* and *sorcerer* interchange-
ably, it's important that we carefully distinguish what was practiced
by the *tonalli*-readers from true sorcery and the darker aspects of
witchcraft, or what's sometimes called *nagualism*. The latter was also
of great importance in traditional Mesoamerican society, but it was far
more secretive and informal by nature, and was associated with negative forces of disease and social conflict. The day priests, by contrast,
worked to help forestall or at least avoid the darker powers that were
perceived to be at work in the universe. The following description from
Sahagún well summarizes the basic attitudes of the time toward soothsayers and priests:

> The soothsayer, the reader of day signs: The soothsayer is a wise
> man, an owner of books (and) of writings. The good soothsayer
> (is) one who reads the day signs for one; who examines, who
> remembers (their meaning). He reads the day signs; he brings
> them to one's attention.
>   The bad (soothsayer) is a deceiver, a mocker, a false speaker,
> a hypocrite—a diabolical, a scandalous speaker. He disturbs,
> confounds, beguiles, deceives others.[7]

The use of the 260-day calendar died out among the Nahua descendants
of the Aztec by the eighteenth century, after hundreds of years of close
regulation by the Catholic Church had banished its use. As one Spanish priest wrote, "It is not a calendar but a soothsaying device in which
are contained a great deal of idolatry, many superstitions, and many in-

vocations to the demons, tacitly and openly."* But even today a vestigial presence of the Aztec *tonalpohualli* appears in a few isolated Nahua communities, in the form of names for spirits and deities still venerated and feared by people from all walks of life. In one village in the Mexican state of Puebla, for example, an important corn spirit is still called Chicomexochitl, "Seven Flower," a name clearly based on the ancient calendar.[8]

Time on a Page

The 260-day calendar was not the only Mesoamerican calendar. As we'll explore in great detail a bit later, the Aztecs, Maya, and others also kept a concurrent tally of days based on the solar year of 365 days. On the first day of the year, the corresponding day in the 260-day cycle was designated as its "year bearer," and could serve as a designation for the entire year as well. Because the 20 days of the *tonalpohualli* cannot go evenly into a year of 365 days, only 4 day signs can ever serve as year bearers. In the Aztec system these were the third, eighth, thirteenth, and eighteenth days, House, Rabbit, Reed, and Flint-knife. This gets to be a bit confusing, especially in Aztec art and writing, which makes heavy use of the year bearer system, but context usually makes clear whether, say, 4 Rabbit or 10 Reed refers to a day or to a year. Both are possible.

This comes into focus a bit more when we consider how the Aztecs, like all other Mesoamericans, saw their calendars as multivalent and layered upon one another. Time was itself an extension of space and of the world directions, all being fused together as different yet complementary aspects of the universe. Probably the most famous representation of this very modern, almost Einsteinian concept of temporal-spatial overlap comes from the divinatory book or almanac known as the Codex Fejervary-Meyer, a *tonalamatl* painted by a master artist living in what is today Veracruz or Puebla, perhaps only a few years before Cortés's arrival.

---

* Sahagún, Florentine Codex, Book 4, p. 181. This quotation comes from a section of the Florentine Codex presenting the pros and cons of stamping out the native calendar.

This single page effectively condenses numerous layers of information regarding the structure of the Aztec and central Mexican cosmos. The overall design resembles a Maltese cross showing four distinct outer sections, each bounded by a different-colored border framing a scene of two gods flanking a world tree. The colors of the frames correspond to the colors that ancient Mesoamericans associated with each of the four quarters: the red frame at the top of the page thus shows east, the left shows north, the bottom is west, and the right shows the southern quadrant. These connected scenes all radiate out from a central square design showing an image of the warrior god Xuihtecuhtli,

Representation of cosmic time, directionality, and centrality; from the Codex Fejervary-Mayer.

the god of heat and fire, and also by extension, an animate of time itself. (His name is composed of *xihuitl,* meaning "turquoise" as well as "year," and *tecuhtli,* "lord.") Streams of blood flow into his body from the four corners, originating outside of the frame from floating body parts of another deity, Tezcatlipoca (Smoking Mirror), the god of darkness and strife, who here appears to have been killed and dismembered, his blood nourishing the victorious Xiuhtecuhtli. The juxtaposition of the two gods, one alive and one sacrificed, may well relate to the victory of light and time over the forces of darkness and night.

At the four corners we see small birds in flight, diving toward the center of the page. Their bodies are day signs from the 260-day calendar: running counterclockwise from upper left are Reed, Flint-knife, House, and Rabbit, making up the four year bearers of the central Mexican calendar. If we look closer still, we see that other individual day signs are placed around the composition, integrated into the corners of the four side frames and in the smaller frames that connect with the corner birds of the year bearers. For example, just to the upper left of the Xuihtecuhtli image we see the toothy reptilian face of the day sign Alligator (*Cipactli*). We see above this, running straight along the frame's line, another day sign, Jaguar, at the upper right corner, and then running left to right above the tree, another day sign, Deer. The same structure exists throughout the composition, so that the frame line decorated with small dots incorporates a variety of day signs, moving around each directional frame, and in and out of the entire design. The dots are key here, for there are twelve small circles in each line connecting the day signs. These represent counts of days, so that twelve days separate Alligator from Jaguar, and twelve more days separate Jaguar from Deer, and so on. The entire design thus represents a counterclockwise representation of the 260-day year, composed of 20 segments of 13 days, all integrated further into a four-part system where each world direction is associated with a set of 65 days. A complete run-through, beginning with Alligator and incorporating the directional year bearers, can best illustrate this structure:

EAST	. . . Alligator + 13 = Jaguar + 13 = Deer + 13 = Flower + 13 = **Reed**
NORTH	+ 13 = Death + 13 = Rain + 13 = Grass + 13 = Snake + 13 = **Flint-knife**
WEST	+ 13 = Monkey + 13 = Lizard + 13 = Earthquake + 13 = Dog + 13 = **House**
SOUTH	+ 13 = Vulture + 13 = Water + 13 = Wind + 13 = Eagle + 13 = **Rabbit**
EAST	+ 13 = Alligator . . .

The temporal scheme on the page represents a complete tally of the 260-day system, grafted on a cosmic "map" where 4 segments of 65 days each are associated with one of the world quarters. Using this page, a Nahua day priest could quickly discern the associations of a given day with a certain direction.

We will occasionally return to this time-space overlap, especially with regard to the year bearers and Mesoamerican concepts of the solar year. The Aztecs and other peoples of central Mexico were very keen on this multidimensional system, although the Maya emphasized similar concepts in somewhat different ways.

The Maya Days

By the time the Aztecs arrived on the Mesoamerican scene, the Maya had already been using the very same 260-day calendar for 1,000 years. The names are similar in many ways, although the antiquity of the Maya calendar makes some of their meanings very hard to analyze, and even obscure in origin. They are far more difficult to translate, therefore, so we regularly refer to them by their Mayan names as they existed in Yucatán in the sixteenth century, preserved for us from various important documents from the colonial period:

1. Imix	11. Chuwen
2. Ik'	12. Eb
3. Ak'bal	13. Ben
4. K'an	14. Ix
5. Chikchan	15. Men
6. Kimi	16. Kib
7. Manik	17. Kaban
8. Lamat	18. Etz'nab
9. Muluk	19. Kawak
10. Ok	20. Ahaw

So, to reiterate the structure of the system, a sequence of days in a Yukatek calendar might look something like this, if picked up in midstream:

... 12 Manik', 13 Lamat, 1 Muluk, 2 Ok, 3 Chuwen, 4 Eb, 5 Ben, 6 Ix, 7 Men, 8 Kib, 9 Kaban, 10 Etz'nab, 11 Kawak, 12 Ahaw, 13 Imix, 1 Ik', 2 Ak'bal, 3 K'an, 4 Chikchan, 5 Kimi, 6 Manik', 7 Lamat, 8 Muluk ...

These Yukatek names are not necessarily the same as those used by other Mayan groups, nor do they necessarily correspond to the names once used by the ancient inhabitants of Palenque or Copan. But we routinely use them out of long-established convention. In fact, other Mayan languages show somewhat different names, which have come down to us from various sources. Among the highland K'iche' Maya of Guatemala, where, incredibly, this 260-day calendar is still in use, the names are different from those used in Yucatán, even if the meanings correspond in most cases, and the resemblance among many of the names is obvious (see Appendix 1). We can be sure that the names in all Mayan languages derive from a much older prototype, which probably even predates the names used by the ancient Maya more than a thousand years ago. And given the variety of languages spoken in ancient times, I suspect that even among Classic-period communities there existed some degree of variation in the names of the individual days, even if the meanings were generally equal across the board.

The Maya manner of writing the days looks quite different from the Aztec method. In shape and form, Maya days conform to a rounded square or rectangular shape, just like all other elements of the hieroglyphic script. In most instances the day signs were placed inside a frame or cartouche, making them easy to spot when appearing in a long string of glyphs in an inscription. The cartouches themselves could be simple ringlike frames, but often they show a three-part scroll, or "pedestal,"

The twenty Maya day signs and their sixteenth-century Yukatek names.
(Drawing by the author)

below. The Maya, like their early contemporaries, opted to write their numbers with the bar-and-dot method, where a single bar represented 5, and a single dot, 1 (with 13, for example, being 2 bars and 3 dots: 10 + 3). The *trecena* number 1 to 13 always was written before or above the day sign, that is, either to the left or on top of the cartouche. This Maya system is a very old one, in all likelihood borrowed from or developed in tandem with that of Zapotec and late Olmec scribes of the Preclassic era.

◇

For the ancient Maya, and among many of their descendants, the day was called *k'in*, a deceptively simple word that also carries the meanings of "sun" and, more abstractly, "time." In some ritual contexts, *k'in* can also be a religious festival, perhaps a reflection of Catholic influence and the importance of saints' days and their associated ceremonies. The ancient hieroglyph for this word seems to represent a four-petaled flower, perhaps a natural symbolic equation of the radiant sun with a bright blossom. The word can also be shown equally well by the face, in profile, of the Maya sun god, K'inich Ajaw, "The Great Sun Lord." K'inich Ajaw is depicted often in Maya art with a large eye and a prominent Roman nose, and with small *k'in* symbols on his face, forehead, and limbs. As a ruler of the heavens, he dresses in jade and cloth, the finery of a Maya king. Often he sits on a throne or inside a radiating cartouche we call a solar disc. K'inich Ajaw, the animate sun and the heat and light of the day, was one of the most important and prominent of all deities in the Maya pantheon.

A wonderful illustration of the 260-day system comes from an ancient wall painting discovered by archaeologists in 1937 at the ruins of Uaxactun, in what is today northern Guatemala. There, while clearing the rooms of a small palatial complex known as Structure B-XIII, they came upon a beautiful wall painting depicting intimate scenes of nobles, musicians, and dancers, and with several individuals sitting in rooms conversing. Judging by its style and the architecture, the archaeologists dated it to about AD 450. Below these scenes of palace life the excavators uncovered a horizontal line of seventy-two painted hieroglyphs, each a different sign in a circle with bar-and-dot numbers (1 through 13) above. Clearly this repre-

K'inich Ajaw, the Maya solar god. (Drawing by the author)

sented a sequence of days in the 260-day calendar—an ancient Maya wall calendar in a literal sense. The days are simply recorded one after another, reading left to right, but a few have short lines of glyphs dangling beneath, evidently marking those days with some descriptive phrase or label; they are very hard to read. Given their context on a room's wall, one wonders if

A sequence of day glyphs discovered written on a sixth-century temple wall at Uax-actun, Guatemala: 7 Imix, 8 Ik', 9 Ak'bal, 10 K'an, 11 Chikchan. (Drawing by the author)

this line of days was used to record the history of the scene above it or, alternatively, was a handy schedule of prescribed rituals for that particular shrine. Sadly, the Uaxactun mural was destroyed by local vandals soon after being exposed by the archaeologists; a few photographs remain for study, along with a watercolor copy made from the original.*

The Maya version of the 260-day round is sometimes called the *tzolk'in*, a term that in Yukatek means "to order the days" (*tzol*, "to order," and *k'in*, "day"). Some writers have objected to the use of the term *tzolk'in*, assuming it's not an authentic Mayan phrase at all, but instead a misnomer coined in the 1920s by the early eccentric Mayanist William Edmond Gates. Even so, *tzolk'in* has been used ever since, if grudgingly.† To me it seems a perfectly good and authentic word, with identical forms well attested in other Mayan and Mesoamerican languages. For instance, a dictionary written in 1940 among the Ch'ortí Maya contains the entry *tzohrk'in*, "a series of days, Christian calendar." (The *r* in Ch'ortí is equivalent to *l* in its sister languages.) In the 1920s, the Jakaltek Maya of highland Guatemala called their ancient 260-day calendar *pisom tsaiik*, "that which sets the days in order." Recall that the Nahuatl term for the same system was *tonalpohualli*, made up of *tonalli*, "day sign," and the verb *pohua*, meaning "to count." The sense is basically the same: a "count or ordering of days." No ancient term for the calendar had so far been recognized in the ancient glyphs, but I'm fairly confident that Gates was right, and that the word used in the courts of Tikal or Copan for the basic 260-day calendar was probably *tzolk'in*, "the

* The Uaxactun paintings were published in Morley, 1946, and in the technical report of the dig by Smith, 1950.

† Gates was a colorful and important character in early Mayan studies, obsessed with tracking down obscure documents and other sources on Mayan languages. He traveled far and wide visiting archives, seeking items that might give clues to the early stages of Mayan languages and their historical connections. Gates was independently wealthy and fronted the money to form The Maya Society in his native Baltimore, along with a series of important publications. He was not an archaeologist, nor even much of an academic, and established scholars of the time saw him as a bit of an eccentric, but he did what few others at the time would do—he rescued many key documents from obscurity and destruction.

ordering of days" or "the lining up of days." The idea is simply conveyed by the Uaxactun painting we've just seen, with its visual presentation of all the days lined up, like an ancient Maya Day-Timer.

Tzolk'in may also convey philosophical understanding of time. The word *tzol* or *chol* in Mayan languages means "to order" or "line up," but it also widely carries the sense of "to interpret," "to read," or "to clarify." For example, in Yukatek, *tzolt'an* (order speech) is "an orator who makes an interpretation, a relator" (*orador que hace algun rasonamiento*). That is, *tzol* is an action, and it structures and gives meaning to something otherwise messy or obscure. One wonders if underlying the term *tzolk'in* is a sense of giving order and reason to the apparent day-to-day randomness in people's experience of time's passing. The word might imply that we are dealing with "ordered" time that is interpretable and understandable, much in the sense that a modern-day diviner or daykeeper facilitates. The 260-day round was the calendar most important in prognostication and divination, acts that themselves give a sense of order and meaning. Indeed, as we will see, divination rites based on the 260-day cycle still play a vital part in the lives of many highland Maya people, even to this day.

The Meanings of the Maya Days

We have yet to explain just what the individual days are all about, and why ancient Mesoamericans chose a sequence of names that include meanings such as "Alligator," "Wind," "Jaguar," and "Rain." Many are words for animals, obviously; others are forces or elements of nature. As the anthropologist Ruth Bunzel observed about K'iche' Mayan day names, "In these twenty sacred words are expressed all of the basic forces of creation and destruction, good and evil, yielding and immutable, operating in the world, in society and in the heart of man."[9] This may well be true, but we ought to remember that the names are of great antiquity, and hearken back to the very beginnings of Meosamerican civilization itself. Over the centuries their original meanings have become opaque, and at times even lost to the soothsayers who use and interpret them.

Here I offer some brief and rather sketchy interpretations of the twenty Maya day names and their corresponding signs, based on their visual forms and considering their original names as we can best reconstruct them. (The names given are the standard Yukatek forms we customarily use, with Classic-era names shown in parentheses along with their probable meanings.)*

1. *Imix* (*Imox*, A Water Serpent). *Imix* or its variant name *Imox* refers to a mythical watery creature we call the Water Serpent, an important animate spirit associated with rivers, lakes, and pools. The head of this fantastic creature can appear as the day sign, but a far more common form of *Imix* is a sign that in other settings stands for *ha'*, "water," in Maya writing. This abstract form, showing an inner circle above a series of short vertical stripes, has its visual origin as a water lily blossom. Elsewhere in Mesoamerica, the first day had similar meanings. Among the Aztecs it was named *Cipactli*, "Alligator."

2. *Ik'* (*Ik'*, Wind, Breath). The name of the second day is consistent in Mayan languages as *Ik'* or *Iq'*, meaning "wind" or "breath." In Nahuatl it was *Ehecatl*, also "wind." The *T*-shape hieroglyph in Maya writing regularly stands for this same word, as in the speaking of the name of the Classic-era Maya "wind god," Ik' Kuh. Visually the *T* within the circle originated as a depiction of a musical gourd-rattle with a small, sonorous *T*-shaped opening in its center. Such maraca-like instruments were often shown being held by the Maya deity associated with music and flowers, perhaps equivalent to the Aztec "Flower Prince," Xochipilli. This connection of wind and breath to music and sound may stem from the Mesoamerican belief in an intrinsic

---

* The discussion of day names and their meanings is based on a variety of sources, and on my own interpretations. For a discussion of Mayan names and their symbolism, see Thompson, *Maya Hieroglyphic Writing: An Introduction*, Carnegie Institution of Washington, 1950, whose text is still extremely valuable, even if his understanding of larger iconography and symbolism is now dated. For Aztec day signs, Alfonso Caso's exhaustive discussion (1967) is also of great use.

connection between wind, the movement of air, and the conveyance of sound. Perhaps for this reason a deified patron of music was the symbol of the day for "wind."

3. *Ak'bal* (*Ak'bal,* **Night**). Its meaning is "night" or "darkness" throughout Mayan languages. The form of the day sign is difficult to interpret and understand, but it may derive from an artistic convention of representing things dark and un-shining. In Classic-era Maya art and iconography, *ak'bal* signs decorate the bodies of certain Underworld deities, including the nocturnal aspect of the sun god. In Postclassic cultures, the meaning was very different, as "house," though perhaps this derived from an idea of the Underworld as a dark, interior architectural space.

4. *K'an* (*K'an,* **Ripe Maize?**). The geometric form of the day sign is the glyph for "maize," and its animate form is of the youthful maize god, sometimes showing corn foliage emerging from the back of his head. The name *K'an* means "yellow," but also, by extension, indicates the color of ripe corn. Other Mayan languages give as a name for this day *K'at,* perhaps meaning "net." Among the Aztecs it was called *Cuetzpallin,* "Lizard." In general, the fourth day is very hard to interpret given its inconsistent names, but visually its Maya sign was derived from a representation of maize.

5. *Chikchan* (*Kan,* **Snake**). The meaning of the fifth day is "snake" in nearly all Mesoamerican languages, including Mayan. Its common glyph is a head, in profile, representing a snake, sometimes displaying sharp frontal fangs.

6. *Kimi* (*Chamel,* **Death**). A skull sign was used to indicate the sixth day in all Mesoamerican scripts, and its name means "death" in nearly all Mesoamerican languages.

7. *Manik'* (*Chij,* **Deer**). The meaning "deer" is conveyed by the visual sign used throughout Mesoamerica, including some Maya forms. The

Yukatek Mayan name *Manik'* is obscure in origin, but might have been borrowed in ancient times from the Zapotec word *mani'*, "animal." In the Classic period the sign usually depicts a curved hand, which in other settings is used to spell the sound *chi*. This came to be used for the day sign because the ancient word for "deer" was *chij*.

8. *Lamat* (*Lambat*, Star?). The visual origin of the eighth day is a representation of a star, with its pointed shape more clearly indicated in early forms. The names for this day present a good deal of confusion, however. The corresponding Aztec day is clear enough as *Tochtli*, "Rabbit," but no Mayan language suggests the same meaning. Instead we find the names *Lamat* or *Lambat*, which make little sense etymologically. Some Mayan languages call this day *Q'anil*, and it is interesting that in the languages of Chiapas, the similar form *k'anal* or *q'anal* is the word for "star."

9. *Muluk* (*Mulu'*, Water Jar?). The ninth day is represented in Maya writing by a ceramic water jar adorned by a diagonal or curved line running down its center. This points to the meaning "water" we find in a number of other Mesoamerican groups, though curiously *not* in Mayan languages. *Muluk* or *Mulu'* is of obscure meaning, but it's interesting that the word *mul* means "water jar" in one Mayan language, Huastec. The original meaning should be clear enough as "water," though perhaps in the original sense of "water jar."

10. *Ok* (*Ook*, Dog). The tenth day means "dog" in much of Mesoamerica (Aztec *Izcuintli*, for example), and the Maya glyph seems to convey the same meaning. It's not clear how the Yukatek name *Ok* relates to "dog," since this word usually means "foot" or "leg" in Maya writing. *Tz'i'*, "Dog," is the name in some Mayan languages of highland Guatemala. In the lowlands during the classic period, the name was probably pronounced *Ok* or *Ook*, though retaining the visual sense of "dog."

11. *Chuwen* (*Batz'*, Monkey). "Monkey" is the meaning of the eleventh day in all Mesoamerican languages. In most Mayan languages

this is *Batz'*, "Howler Monkey." The Maya day sign shows a howler monkey, which was also a mythological symbol associated with scribal arts and creativity. The name *Chuwen*, also meaning "artist," refers to this connection.

12. ***Eb (Eb,* "Tooth").** Early visual forms of the glyph for the twelfth day emphasize teeth on a fleshless jawbone. The Mayan names *Eb* or *Eb*, depending on the language, probably derive from words for "tooth," *e'* or *eeb*. In later examples of the Maya glyph, the jawbone symbol was elaborated into a skull-like form, but still looking different from the generic skull of the sixth day, *Kimi* or *Chame*.

13. ***Ben (Been,* Reed?).** The thirteenth day means "reed" in most Mayan languages, and this was probably the visual origin of the Maya day sign. The Yukatek name *Ben* or *Been* has unknown meaning, but in highland Mayan languages its name is *Aj*, "Reed."

14. ***Ix (Hix,* Jaguar).** The day is "Jaguar" throughout Mesoamerican languages. The common word for "jaguar" in Mayan languages is *bahlam*, but *hiix* was evidently a variant term also used for large spotted cats.

15. ***Men (Tz'ikin,* Bird).** The fifteenth day is "Eagle" in most Mesoamerican languages, although the highland Mayan name *Tzik'in*, meaning "bird," is more generic. The Maya day sign seems to represent the head of a mythological avian being known as the "Principal Bird Deity," an important Creator being who had four directional aspects. The name *Men*, found in Yukatek, could relate to the word meaning "to make, create."

16. ***Kib (Chabin,* ?).** In terms of meaning this is probably the most obscure of the Maya days. In much of Mesoamerica this day is "Vulture," but among the Maya its significance is far less certain. In Yukatek, *kib* is a word for "beeswax," but this is difficult to relate to the ancient glyph. Visually, as well, the day sign is hard to make out.

17. *Kaban* **(*Kaban,* Earthquake?)**. Its glyph corresponds exactly to the sign otherwise meaning "earth," *kab,* and this seems to have been its basic meaning among the Maya. It may have had a more specific sense of "earthquake," as suggested by its names in other Mesoamerican languages, including Nahuatl: *Ollin,* "Earthquake."

18. *Etz'nab* **(*Chinax*[?], Knife)**. The sign originated as a representation of a flint blade or knife, its basic meaning throughout Mesoamerican cultures (Aztec: *Tecpatl,* "Flint-knife"). The ancient name of the Maya day is difficult to know; in highland Mayan languages it is *Chinax* or *Tijax.*

19. *Kawak* **(*Chahuk,* Lightning, Storm)**. The nineteenth day means "storm" or "lightning." Its animate head glyph is a representation of the deity Chaak or Chahk, whose name is in fact related to Mayan words for "storm." The simplified Kawak sign is also used in Maya script as the word for "stone" (*tuun*), perhaps reflecting an intimate association between rain and sacred stones and caves.

20. *Ahaw* **(*Ajaw,* Lord)**. The final Maya day means "lord" or "king." It's by far the most common of Maya day signs, simply due to the fact that many important calendar festivals fell on Ahaw, corresponding to the so-called period endings of a much longer and more complex calendar, which we will discuss in the next chapter. It can be represented as a rather simple face-like design, or as the profile head of a man wearing a headscarf. This day meaning "king" thus "ruled" over particular segments of time, much as a ruler reigned over a court and a kingdom.

Such is my superficial attempt to make sense of the twenty ancient day names. It's important to realize, however, that the meanings of days as they passed in real time depended on many factors, including their associated numbers. As experienced, then, the divinations for particular days may well have been different depending on time and place. This is certainly true in areas where the calendar survives today, where curers and daykeepers provide varied meanings and interpretations of the days.* In this sense the meaning of time can be very localized.

Time on a Plate

Much as we saw with the Aztec integration of space-time on the page of the Codex Fejervary-Meyer, the sequence of twenty Maya day signs had its own internal arrangement that reflected cosmological and spatial associations. One artist depicted this idea in a similar but special way on a painted dish or plate that was probably used for ceremonial occasions at a Classic-era Maya court. Taking advantage of the circular form of the plate, or *lak*, the painter depicted a wheel of sorts, with concentric rings surrounding a stylized image of a maize god, shown covered in jade jewels and a quetzal feather decoration, both evoking the green shimmering leaves of the corn plant. His arms are extended in a gesture that suggests a ritual dance, not unlike those that lords and nobles performed in their plazas and courtyards.

Around the maize god we see a ring of twenty hieroglyphs divided into four segments, each oriented to one of the four world directions. These are the twenty days of the *tzolk'in*, arranged in their proper sequence but parsed in a somewhat unusual way, in groups of five signs. For example, we see just above the head of the maize god the sign for the seventeenth day *Kaban* (Earthquake), with the next four day signs. Running clockwise, their order is (here indicating their day number):

17. *Kaban*, 18. *Etz'nab*, 19. *Kawak*, 20. *Ahaw*, 1. *Imix*,
2. *Ik'*, 3. *Ak'bal*, 4. *K'an*, 5. *Chikchan*, 6. *Kimi*,
7. *Manik'*, 8. *Lamat*, 9. *Muluk*, 10. *Ok*, 11. *Chuwen*,
12. *Eb'*, 13. *Been*, 14. *Ix*, 15. *Men*, 16. *Kib*

* Space limitations prevent any detailed account of the divining methods used by modern ritualists, but several excellent studies describe the process, focusing on different highland Guatemalan communities. See for example Bunzel, 1959; Colby and Colby, 1981; and B. Tedlock, 1982.

Representation of Maya days around the Maize God, from a painted plate.
(Drawing by the author)

The first day sign in each segment—days 17, 2, 7, and 12—corresponds to the year bearer days used in the Classic Maya system, or one of the four days on which the 365-day solar year could begin. The layout of the plate implies that each "string" of five days was associated with one particular direction.

Why would the maize god be featured on a plate like this, surrounded by symbols of time? Well, a plate such as this would have been used as a container for a variety of tasty delicacies enjoyed by elite members of Maya society, especially the ubiquitous tamales (*waaj*) stuffed with fish, deer, or iguana meat. The maize god, as the essence of corn, marks the center of the plate as a place for food and sustenance. Time frames the deity as if it is in some way an extension of him—an idea that rings true when we consider the cyclical nature of corn's growth and harvesting over the course of a year. The maize god was closely associated, too, with the animate earth, and we see in this plate how the day "Earth" hovers over the maize god's head, almost as if it is a label. The ring of twenty days seems distantly related, as we will see, to the layout of the famous Aztec Calendar Stone, with its circular display that included twenty days around the central image of the Earth Lord.

The Survival of Maya Timekeeping

It's no small miracle that this very same 260-day calendar is still carefully kept in a few remote areas of Mexico and Guatemala, where it continues in some places to serve as a very public and vital part of community ceremonialism. This is especially true in highland Guatemala, among Maya peoples who have successfully guarded their cultural knowledge over the last five centuries. The system of day reckoning is essentially the same as that used by the Aztecs and Maya in the fifteenth century, representing the most important and incredible continuity between the pre-Columbian past and the modern world.

The remarkable survival of the calendar became known only after early researchers in the United States and Europe began to unravel the inner workings of the ancient Maya calendar using only historical documents and hieroglyphic manuscripts. The Maya 260-day calendar remained unknown to the Western scholarly world until 1841, when the writer and explorer John Lloyd Stephens published a summary of the work of Juan Pío Pérez, the historian he had befriended in Mérida, Mexico, in the course of his famous travels. Pío Perez had accumulated and studied a number of native documents written in Yukatek Mayan, many containing detailed historical records that referenced the ancient calendar systems. Many of these documents were copied and recopied over several generations by careful village scribes, and some might even have been copied from pre-Columbian screen-fold manuscripts. Much if not all of the calendar was forgotten by the Maya who lived in nineteenth-century Yucatán, yet the priests and scribes had kept the ancient records, the *uchben tz'ib*, close at hand. Pío Perez studied the many versions of these native chronicles and was able to see the mechanics of the 260-day cycle. No one in the 1840s knew the hieroglyphs for the days, but Pío Perez saw the structure, and Stephens recognized its importance.[*] Ironically, Pío Perez, Stephens, and many other nineteenth-century scholars were not aware that the 260-day calendar was still actively in use in many villages in the highlands of Guatemala and in Chiapas, as anthropologists would find out years later.

A watershed moment came in 1863, when a French priest named Charles Brasseur de Bourbourg discovered an old manuscript in the archive of the Royal Library in Madrid. This changed everything, for it was a copy of a book called the *Relación de las Cosas de Yucatán* (Relation of the Things of Yucatán), written by Friar Diego de Landa, bishop of Yucatán, in the sixteenth century, just scant decades after the conquest of that region. Landa's work is a gold mine of information about the life, religion, and history of the region on the eve of European contact—a veritable Bible of Maya studies. In his own day, Landa had gained a harsh reputation for his overzealous reaction to "idolatry" and for stamping out native "works of the devil," including the burning of hundreds of hieroglyphic books. Yet his manuscript is replete with detailed information on Yukatek Maya history, culture, and customs. Lucky for us, Landa's work included images of the "letters or characters" for each of the twenty days. With these drawings in hand, it was not difficult for early scholars to match the day signs with those in the surviving ancient Maya books and then, ultimately, to the calendar records in the older stone inscriptions from Copan and Palenque, accurate reproductions of which were being published and circulated by the 1880s.

---

[*] Stephens's 1841 book was the second of his famous works documenting his explorations among the ruins of what is now Mexico, Guatemala, and Honduras (Stephens, 1839, 1841). These two works, accompanied by the accurate and evocative drawings of Frederick Catherwood, marked the beginnings of serious scholarly interest in the ancient Maya. Often overlooked, Juan Pío Perez deserves equal billing as one of the founding fathers of Mayan calendar research and scholarship.

With Brasseur's publication of Landa's work, interest in Maya studies saw fervent growth in academic circles, and provided all of the
necessary groundwork for the even more important advances we will
discuss further on. Other historical documents stored away in remote
archives began to be copied and studied. In 1877 a Guatemalan named
Juan Gavarrete found an old manuscript of a native calendar, which
had been collected in the 1700s by the archbishop Don Pedro Cortés y
Larraz, containing a description of the calendar and lists of days and
their prognostications. The manuscript was called the Calendar of the
Indians of Guatemala, 1722, written in the K'iche' Mayan language.
The archbishop had found the calendar during one of his sojourns into
the Guatemalan countryside, perhaps in the town of Quetzaltenango.
He and other Catholic authorities took great interest in the way the native daykeeping was being used to schedule masses and other religious
events, but they nonetheless eradicated the tradition whenever possible, and several such Indian calendars were confiscated and archived.
Once Gavarrete had found and copied the manuscript, it sat unstudied
by other scholars; incredibly, it didn't see the light of day until well into
the twentieth century.

What came as a surprise to scholars in the late nineteenth and early
twentieth centuries were direct encounters with the ancient calendar.
As early as 1888, a Mexican writer named Vicente Pineda published a
list of day names then used by the Tzeltal Maya of Chiapas, although
his revealing work, published in remote southern Mexico, remained
obscure to many of the leading scholars of the time in North America
and Europe.[10] Unfortunately, the use of the calendar among the Tzeltal
was soon lost, perhaps in the early twentieth century. It wasn't until
the 1920s that the existence of these survivals became widely known
as anthropologists probed the remote mountainous terrain of what is
now southwestern Guatemala, interviewing local Maya priests and
civil authorities.

One of these intrepid field workers was the noted writer and anthropologist Oliver La Farge, who established a firm reputation as a
linguist and ethnographer at the same time he was publishing novels,

From left to right: Oliver La Farge, Frans Blom, and assistant Lázaro Hernández during their 1925 Tulane expedition through Mexico and Guatemala.

even winning a Pulitzer Prize for his work *Laughing Boy* in 1930.* In 1925, La Farge had accompanied an important expedition to southern Mexico and Guatemala, exploring the Isthmus of Tehuantepec and the nearby region of Chiapas in the company of Tulane archaeologist Frans Blom. The two men had already made remarkable finds, including the initial documentation of the great Olmec ruins of La Venta, in the state of Tabasco. One day, after weeks of difficult travel, they found themselves passing time in the plaza of Comitan, a large town in southern

---

* The La Farge family had other notable members. Oliver La Farge was grandson of celebrated American artist and designer John La Farge (1835–1910); Oliver's son Peter La Farge (1931–1965) was a singer and a well-known member of the Greenwich Village folk revival in early 1960's New York City, collaborating with Pete Seeger and the young Bob Dylan.

Mexico near the border with Guatemala. They were speaking with one of Comitan's well-known citizens, a Don Gregorio de la Vega, about the Indian communities of the area and in Guatemala, communities that, in those early days of Maya field ethnology, had hardly been visited or studied by outsiders. Don Gregorio had traveled among many Indian towns more than most other Ladinos, and he mentioned to Blom and La Farge his experiences in a place across the border called Jacaltenango, where he said the people held a ceremony devoted to something called "year bearers." As Blom later wrote, "Our informant probably did not realize how deeply we were thrilled by this casual remark." Amazed at the prospect of seeing the ancient calendar and its associated rituals still in action among living people, speakers of a Mayan language, the two men lost little time in making their way to Jacaltenango the next week. There they were quickly able to confirm what Don Gregorio had told them. The Maya calendar, in some parts at least, was alive and well, if still kept largely under wraps. As La Farge himself wrote, somewhat in awe, "It appears that the shamans, many of them for generations having had no writing, have maintained their count of days unbroken and without error since the time of the conquest."[11]

Plans were made for a thorough and in-depth investigation, calling for La Farge to return to Jacaltenango with an assistant, Douglas Byers. They went back to the region in 1927, and stayed for months interviewing the local calendar priests. In his letter of presentation to La Farge and Byers's classic book on Jacaltenango, *The Year Bearer's People*, Blom admiringly writes:

> Few people realize the task before these two men. Their objective was to investigate the ancient and secretive religious ceremonial of a tribe of Indians which had been persecuted by well-meaning missionaries for 400 years. They were to conquer the suspicion of the Indians and to gain their confidence. Their task can be linked to that of a man trying to become familiar with the ritual of a Masonic Lodge without becoming a Mason himself. Outwardly there was nothing spectacular about their work. They would have to spend long, monotonous days

wandering around among the natives in hopes of finding a clue about some ancient ritual.[12]

Their discovery and documentation of the use of the calendar in its social and religious context opened many eyes at the time, and helped to bridge the gap between Maya archaeology and ethnology—two fields that have mutually benefited from an important synergy ever since. As the celebrated Mayanist J. Eric S. Thompson wrote in the 1970s, "Less than fifty years ago no one dreamed that some Maya communities not only retained the old sacred almanac of 260 days, but were still regulating their lives by it."[13]

Within a few short years more living calendars came to the attention of the academic world, though it's probably correct to say that these native systems of timekeeping had been fairly out in the open, if anthropologists had only looked for them.* On the heels of La Farge and Byers's studies in the Jacaltenango region, others descended upon the various communities of the Guatemala highlands to study and document what they could find, hoping that some villages had preserved more elements of the ancient calendar than others. One of the most important of these eager travelers was Jackson Lincoln, an anthropologist who had been born into a prominent New York family, and whose experiences in the American West, just as with Oliver La Farge, engendered a keen interest in Native American culture. Lincoln decided to go into an area even more remote than where La Farge had been, among various towns where Ixil Mayan was spoken. There he, too, found numerous daykeepers and curers who made full use of the 260-day calendar. He was able to compare the day names of different towns, and compare their meanings as understood by individual curers. He was delighted to find that the calendars of Ixil towns agreed perfectly with the calendars in the Jakaltek towns La Farge and Byers had visited. Unfortunately, Lincoln caught pneumonia in the midst of

* One likely factor in the general ignorance of the survival of native calendars was the distraction of World War I, and the profound bias of much Mayan research before the 1920s toward Yucatán.

his work among the Ixil. His wife rushed him to a hospital in distant Guatemala City, where he soon died. Some of his notes and observations were published, but the bulk of his important research remains unknown to this day, even to many Mayanist scholars.

Among the remarkable things learned through the study of modern Maya calendars is how the days themselves appear as living entities. "The twenty days are regarded as gods," wrote anthropologist Maud Oakes, in her study of the Guatemalan town of Todos Santos.[14] Each day is "addressed" just as if it is an elder person. The same notion of animate time is found in many places throughout Mesoamerica, in fact.* Living time still exerts a keen influence on the events and lives in traditional communities, much as it has for thousands of years.

Within a short time after the pioneering initial work of La Farge and Lincoln, anthropologists were able to document the widespread existence of the 260-day calendar still in use throughout highland Guatemala and into parts of Mexico. Not only were some Maya groups still making use of it, often in a very public fashion, but so, too, were isolated communities in Oaxaca, speakers of the Mixe language. The calendar's survival is all the more remarkable given the very recent violent history of the Ixil region, where during the 1980s the civil war between the Guatemalan government and insurgents led to the attempted eradication of whole Maya villages through targeted attacks and massacres. Despite this continued pressure from the days of the conquest, this most ancient 260-day divinatory calendar, probably in use since Olmec times, still continues its march forward.

The 260-day Question

Why a cycle of 260 days? Scholars have pondered this question for over a century, ever since Mesoamerican studies began as a serious

---

* For the concept of Mesoamerican days as living entities, see also La Farge and Byers, 1931, pp. 172–73; LaFarge, 1947, p. 171; B. Tedlock, 1982, p. 107; Lipp, 1991, p. 67.

academic discipline. Several different theories have been proposed since that time, only to be rejected and then later resurrected, probably because no single answer is clear-cut or obvious. One possible explanation is simply numerological, since, as we've seen, the two meaningful factors 13 and 20 are both extremely important in Mesoamerican cosmology. The number 20 is closely associated with the human body, with its 20 digits. The number 13 plays a key role, we will see, in Maya numerology and calendrical reckoning, not just in the 260-day cycle. It is probably significant that in Maya hieroglyphic script, separate "gods" can represent the numbers 1 through 13, with 14 being the first exclusively "hybrid" deity incorporating features of 10 and 4. This may reflect a very old religious significance to 13, it being the highest in a set order of numbers that were somehow sacred or divine. Moreover, according to some colonial sources, 13 is the number of layers or segments within the celestial realm. The number 260 thus may interweave notions of the human body (20) with the cosmos at large (13).

However, evidence from historical and modern sources suggests that the meaning of 260 is not just about numbers, and that it may be firmly rooted in the circumstances of human life and experience. As the early Mayanist Charles Bowditch first proposed a century ago, it so happens that 260 days equals about 9 lunar months, or about the span of human gestation.[15] But not all scholars accepted this view. One prominent critic was J. Eric S. Thompson, the Englishman who dominated Maya studies throughout much of the twentieth century: "It has been suggested that the number [260] was chosen because it approximates the period of human gestation, but that is not a very happy explanation because there is no logical reason why the period of pregnancy should be considered in establishing a divinatory almanac."[16] Thompson's inability to see the "logic" of a connection between pregnancy and divination is hard to fathom, but his was a highly influential voice in Mayanist circles, and the debate has continued ever since.

Sometime after Bowditch's initial proposal, the great German ethnographer Leonhard Schultze-Jena studied the 260-day calendar that was (and still is) in use among the K'iché' Maya of highland Guatemala.

We will look at this remarkable survival more closely in a moment, but suffice it to say for now that the K'iché' daykeepers told Schultze-Jena, "with a look suggesting this should have been self-evident," precisely what Bowditch had earlier reasoned: that the 260-day calendar "had to do with the same number of months that are given for a woman's pregnancy."[17] To this day, other traditional calendar priests make the very same observation.

As we will see in more detail later in this chapter, several descriptions of the 260-day calendar from early colonial sources, many written by native authors, emphasize its importance in the divination of a person's fate, based on the timing of childbirth and pregnancy. So, I for one am on the side of Bowditch and Schultze-Jena, and feel that no debate need continue: the 260-day calendar derives its divinatory importance largely *because* it is a reflection of a fundamental biological cycle, uniting human experience with the two fundamental numbers, 13 and 20, that give time and the cosmos much of its structure.

But this may not be the whole story when it comes to explaining the 260-day cycle. Another intriguing explanation for its importance is astronomical. Early in the twentieth century, around the time the human gestation explanation was first being considered, Zelia Nuttall noted that 260 days was the span of time between the sun's zenith passages along the latitude of 15 degrees north, far south within the Mesoamerican region, about where the sites of Copan and Izapa are located. One of these zenith passages corresponds to the day August 13 in the solar calendar, which we will find in later discussions to be a phenomenally important date in Maya time reckoning overall. We know that astronomers past and present pay close attention to the zenith passage, but its possible relation to the 260-day cycle is difficult if not impossible to confirm.*

---

 * The zenith passage model for the 260-day calendar was first offered by Zelia Nuttall, 1927, and most strongly advanced by Vincent Malmström, 1973, 1997. One point in disfavor of the theory is Malmström's presumption that Copan was "the principal center of astronomical studies" for the ancient Maya (see Henderson et al., 1974, p. 543). Today we know that this was not the case. He was apparently

With such varied accounts and explanations, we're left with a difficult, and maybe false, choice. I am convinced that Schultze-Jena's proposal about the period of human gestation is basically correct, given how important this is among the *ajk'in* priests of modern times, whose job it is to keep track of the days and their meanings. I'm also partially convinced that Nuttall and Malmström were on to something about the zenith passages and their connection to 260 days along the 15-degree latitude. It would not at all surprise me to learn, if one ever could know for sure, that both factors were noticed and integrated by early Mesoamerican priests. I'm sure those priests were able to discern the apparent correspondence between human gestation and the zenith interval, even if the connection was not always mathematically exact.

The Solar Year

Whereas the 260-day *tzolk'in* count was geared mostly to divinatory reckoning, another calendar that ran concurrently was based on a number far more familiar to us: the 365-day solar year. Called a *ha'b* in most Mayan languages, this (more or less) "true" year provided a framework for communal agricultural festivals and ceremonies, and served to complement the more private and esoteric nature of the 260-day round. This solar year is still reckoned in the ancient manner in several modern-day Maya communities as a type of "civil calendar," even in areas where the more esoteric 260-day cycle no longer survives.[18] Today, the K'iché' Maya of Momostenango, Guatemala, refer to this also as the *masewal q'ij,* "the common days."[19] When used together, as was usually the case, in ancient times at least, both of these

_____

basing this claim on the antiquated view, typical of the early years of Mayan studies, that a prominent sculptured scene at Copan, on Altar Q, represented a meeting of astronomers. Today we understand that it is primarily a historical monument portraying the sixteen kings of Copan's ruling dynasty (see chapter 9 of this book for a fuller discussion of the altar).

A two-part Calendar Round date, written as 7 K'an 2 Woh, from a tablet discovered at La Corona, Guatemala. (Photograph by the author)

day-reckoning systems made up a larger cycle known to us as the "Calendar Round," lasting 18,980 days, or about 52 years.

The Mesoamerican solar year calendar had 18 named "months," each 20 days in length, followed by a shorter 5-day period and the very end, before the start of a new year (i.e., [18 x 20] + 5 = 365). Individual days would be numbered sequentially within each 20-day segment, so that a day might fall on "the eighth of Mol," to be followed by "the ninth of Mol," and so on, up to 19 Mol (using an Arabic number as a convenient shorthand to represent the station). The next day, in turn, was the "seating" or "base" (*chum*) of the next month, named in this case *Ch'en* in Yukatek terminology. Then came 1 Ch'en, 2 Ch'en, and so on, up to 19 Ch'en, which was then followed in turn by the "seating of Yax." The final five days that made up the last mini-month represented a time of great trepidation and concern. The Maya of Yucatán referred to these days by the name *Wayeb*, but also as the *xmak'aba'k'in*, "the un-named days."

It is interesting to consider why "seating" was used to describe the installation of a new month, for this was the basic term commonly applied to the accession of a king. Perhaps in some real sense the months were not simply abstract subdivisions of the year, but also animate beings in their own right. It is tempting to wonder in this light if the twenty-day period was somehow conceived as a *winik*, a word that means both "twenty" and "man, person" in some Mayan languages. The word for "month" in Yukatek, which is sometimes applied to these periods, is the cognate form *winal*.

All cultures of Mesoamerica made use of this type of calendar, although the individual names of the months varied a great deal among language groups, cultures, and even communities. These were basically names for important community festivals held in a set sequence throughout the year. Early historical sources such as Landa's *Relación* give a detailed sense of how each of these periods involved regular community-wide rituals laden with symbolic rites geared to specific gods. His account of the festival of the initial month *Pop* serves as a good example:

> The first day of Pop, which is the first month of the Indians, was their new year and was a very solemn festival among them; as it was universal and all took part in it and so the whole town jointly made the feast to all the idols. To celebrate it with more solemnity they renewed on this day all of the objects which they made use of, such as plates, vessels, stools, mats and old clothes and the stuffs with which they wrapped up their idols. They swept out their houses, and the sweepings and the old utensils they threw out in the waste heap outside the town; and no one, even were he in need of it, touched it.
>
> . . . For this festival, the lords and the priest, and the principal people began to fast and to abstain from their wives . . . for some set about it three months in advance, others only two, and still others as long as seemed good to them, but no one less than thirteen days . . . Those who once began these fastings did not dare to break them, because they believed some calamity to themselves or to their houses would befall them.*

---

* The description of the festival of *Pop* is from Landa, *Relación de las Cosas de Yucatán*, pp. 151–53.

As with the day names of the *tzolk'in*, Mayanists long ago adopted the Yukatek names when referring to these months, basing the terms on those provided by Landa and other colonial manuscripts. But these terms vary considerably among all Mayan languages, and they do not necessarily reflect the names we find spelled in the ancient hieroglyphic inscriptions. Among the Tzotzil of highland Chiapas, for example, the month names show only the faintest correspondence to those used in early Yucatán. Generally it seems the month names were susceptible to far more change and variability than the day names, even among peoples speaking similar languages. In Classic times as well, there is good indication of some regional variability in the month names, from kingdom to kingdom.

Here are the Yukatek month names and their corresponding glyphs from the Classic-era inscriptions. The likely ancient names of the months are given in the right-hand column, as best as they can be reconstructed:

Glyphs for the Maya "months" with their common Yukatek names.
(Drawings by the author)

YUKATEK	CLASSIC MAYAN
1. *Pop*	*K'anjalaw*
2. *Woh*	*Chakat/Wooh*
3. *Sip*	*Ik'at*
4. *Sotz'*	*Suutz'*
5. *Tzek*	*Kasew*
6. *Xul*	*Tzikin*
7. *Yaxk'in*	*Yaxk'in*
8. *Mol*	*Mol*
9. *Ch'en*	*Ik'sihoom*
10. *Yax*	*Yaxsihoom*
11. *Sak*	*Saksihoom*
12. *Keh*	*Chaksihoom*
13. *Mak*	*Mak*
14. *K'ank'in*	*Uniw/K'ank'in*
15. *Muwan*	*Muwaan*
16. *Pax*	*Paax*
17. *K'ayab*	*K'anasiiy*
18. *Kumk'u*	*Hulohl*(?)
19. *Wayeb*	*Ti'wayhaab*(?) (final five-day period)

The conventional Yukatek names are semantically obscure in many places, but if we analyze the ancient names used in the classic inscriptions, we find that many refer to food plants, seasons, or other telling characteristics of a particular year. *Yaxk'in* is "dry season" or "winter" (literally "new sun"). *Mol* means "harvest." The names *Kasew* and four others incorporating the word *sihoom* (the prefixes are color terms) refer to certain fruits. Each culture throughout Mesoamerica made use of the same calendar structure, but each chose to give varied names to the individual months, reflecting local environments and different agricultural schedules and activities.*

---

* The Aztec 365-day calendar shows a paired format to many of its festival/month names, with one being a "great" version of the month that came before it. We don't see this pattern in the Classic-era Mayan month names, where instead each name and

The lack of a leap year in this system of 365 whole days led to an inherent "drift" of any one station in the calendar with regard to the true solar year of 365.24 days. This means that the day 1 Pop, falling on, say, January 25 in one year, will fall on January 24 after four years, and on January 23 after four more years. With the passage of fifty years, 1 Pop would appear on January 13, and in a hundred years, on December 31. In this way the Mesoamerican solar year, while astronomically based, is not an *observed* solar calendar. The Maya, Aztecs, and others were well aware of the more precise solar year as fixed by the solstices and equinoxes, but for whatever reason, they did not formally synchronize those phenomena into this particular conception of "year-ness," with its basis in whole units of days.

The New Year

I'm writing these words on a very special occasion. February 22, 2009, is the Maya New Year, according to the calendar still widely used in the highlands of Guatemala. Today is the first day of the 365-day calendar, and it falls on 10 Ik' (10 Wind) of the *tzolk'in*, now established as the year bearer. Hundreds of people are gathering in a park here in Guatemala City to celebrate, including indigenous Maya priests from distant communities. Such a gathering would have been impossible or

---

hieroglyph is very different from the others. However, a similar dualistic arrangement may well have existed anyway, subtly evident in hieroglyphic date records that specify a so-called patron of each month. These month patrons preserve the paired arrangement, and were perhaps adopted from early Isthmian or epi-Olmec cultures to the west, from whom the early Maya evidently borrowed the so-called Long Count calendar. The system is very old, in all likelihood, perhaps even having an origin in a now-lost Olmec system. The Mayan pairings of month patrons include "Sky" (for *Tzek*) and "Earth" (*Xul*), "Day" (*Yaxk'in*) and "Night" (*Mol*), "Moon" (*Ch'en*), and "Star" (*Yax*), and perhaps others not so easily discernable. Interestingly, in the Aztec system of the months, if we collapse the paired names, we find that the total number of festivals is no longer eighteen but thirteen, the most sacred of Mesoamerican calendrical numbers. See Rice, 2007, p. 60.

unheard of fifteen or twenty years ago, when, during that country's long civil war, the Maya were hardly out in the open, most caught in the middle of political forces and issues they didn't understand or much care for. Until only recently, the indigenous culture was still largely suppressed or actively hidden away in remote corners of the highlands, some not far from the ruins of the ancient capitals where kings once ruled and warred with one another. Today, thankfully, Guatemala is at peace, and is a very different place. It's a nation fraught with major challenges, to be sure, but it is struggling to make pluralism work, with the Maya now taking an active and important role in shaping the country's course.

One of these great Maya capitals was Iximché, the capital of the Cakchiquel kingdom. It rests on a beautiful windswept plateau high in the cool, pine-covered mountains above the modern town of Tecpan. In 1524 Iximché fell to the Spanish conqueror Pedro de Alvarado, who ravaged the people and surrounding countryside, setting in motion the *violencia* that has beleaguered the region ever since. Despite the sad history of the place, the evocative temples and atmosphere of Iximché make it one of my favorite locales in the Maya world. Often when there, I find hidden in the brush, near one of the ancient temples, a number of small ceremonial shrines, some small and others large. Their burnt candle wax and offerings of flowers and food remind me of those childhood experiences I had at Coba, in distant Yucatán, as a participant in the rain ceremony for the Chaaks. The makeshift altars at Iximché are similar focal points for praying to the gods and ancestors, and to the ancient day lords as well.

Now, on the Maya New Year's Day, I'm happy to learn that Iximché is the site of a major celebration for 10 Ik', the new year bearer, with hundreds of Maya expected to celebrate among the not-so-ancient ruins. Pedro de Alvarado may have conquered this once-great city, but the day lords still rule. Not Alvarado—not anyone—could ever conquer time itself.

6

LONG COUNTING

*Four four-hundreds of years and fifteen score years was the end of
their lives, because they knew the measure of their days.*
—The Chilam Balam of Chumayel

The oldest known American book resides in the archives of the
Saxon State Library, or *Sächsische Landesbibliothek*, in Dresden,
Germany. It's not a printed volume, nor a collection of manuscript pages bound in vellum, written by the hand of some early Spanish friar or explorer. Rather, it's a Maya book about eight centuries old, containing page after page of carefully painted divinatory almanacs and mathematical tables. We call it the Dresden Codex, and it consists of seventy-four beautiful pages painted by a handful of artists and scribes who lived in Yucatán in the twelfth or thirteenth century. Like other Mesoamerican documents that survived the Conquest, the Dresden Codex has a screen-fold format, with each of its pages attached end to end and collapsible like an accordion. The colorful reds, yellows, and blues still grace its plaster-coated pages and remain vibrant to this day, as do the myriad hieroglyphs and portraits of gods. In addition to being a remarkable expression of religious and scientific thought, the Dresden Codex is a stunning work of scribal art, and offers a small hint of the thousands of similar literary works, now lost,

Ernst Förstemann (1822–1906) decipherer of the Long Count calendar.

that once existed throughout the ancient Maya world.

Before 1880, few scholars paid much attention to the strange artifact. Johann Christian Götze, an early director of the Royal Library, obtained it from a private Viennese collection in 1739, and donated it to his institution a few years later. But its earlier origin had always been cloaked in obscurity. Those few who did see and study the book called it Aztec. The document sat quietly in storage until 1880, when a librarian named Ernst Förstemann noticed it and became increasingly fascinated by its mysterious contents. Förstemann kept it close at hand (in his desk drawer!), analyzing and pondering its mysteries in his spare time. He quickly made a number of key observations about the book's numerology and mathematics, and soon his observations led to the first detailed understanding of the Maya calendar—a much-needed kick start that initiated a string of breakthroughs that eventually would lead to the decipherment of Maya hieroglyphs.[*]

Many pages of the Dresden Codex convey calendrical or astronomical knowledge of one kind or another, much of it related to ceremonies associated with particular gods, including the rain deity Chaak, the young maize god, and the moon goddess. Sections of it were used by ancient priests and daykeepers, or *aj k'inoob*, in order to track the movements of Venus and other planets, and mostly to ensure that important rituals for various gods were properly timed and performed. Given the

---

[*] For more on Förstemann's research on the Dresden Codex, see G. Stuart, 1992, and Coe, 1992. English translations of Förstemann's writings on the Dresden Codex and the calendar appear in Bowditch, 1904.

wealth of calendar-related content, it isn't surprising that these codices have proved instrumental in the modern decoding of the Maya calendar and its associated numerology. Many of their pages still defy much understanding, but it's clear that they have already given us a unique window into the arcane details of Maya religious belief and practice.

To me it's truly astonishing that any of these ancient Maya books could have survived the Conquest and its violent aftermath. In the first decades after the conquest of Yucatán, for example, native books containing "lies of the devil" were routinely sought out by the Spanish and destroyed. For this reason, Maya shamans and healers came to closely guard many old books that preserved native history and culture, such as the so-called Books of Chilam Balam. The story of the Dresden Codex's journey from Mesoamerica to Europe remains unclear, but it was undoubtedly a risky and adventurous one. We do know it was documented as being in private hands in Vienna as early as 1739. The Vienna connection is interesting, given that Austria was a part of the Holy Roman Empire ruled by Charles V, who was also (as Charles I) ruler of the new Spanish Empire in the Indies. Some believe that the Dresden Codex and other famous pre-Columbian treasures were in fact collected by Hernán Cortés himself on his initial trip around the Gulf Coast of Yucatán and Mexico in 1519, and sent back to Charles I as part of the "Royal Fifth."[1] In any event, after the mid-eighteenth century the precious manuscript was safely housed in the Dresden library—that is, until 1945, when the horrific firebombing of Dresden by Allied aircraft resulted in significant water damage to its pages. Yet it somehow survived. Given the extent of devastation in the city—some estimate that twenty-five thousand citizens of Dresden were killed—the endurance of this ancient, fragile book is miraculous.

Open the Dresden Codex to pages fifty-eight and fifty-nine and you'll see an array of small numbers—bars and dots—arranged in vertical columns, intermingled with various hieroglyphs and religious imagery. They, too, record dates, but in a very different way from what we've seen thus far in our exploration of 260- and 365-day cycles. Such number sequences usually consist of five places, read from top to bottom in a column, and they typically precede a Calendar Round re-

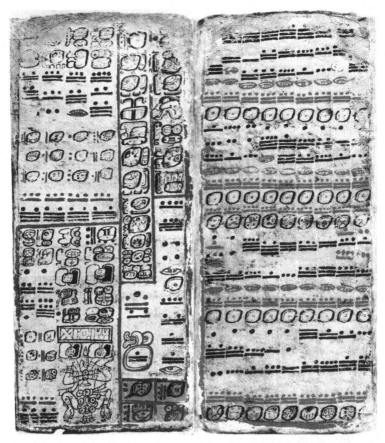

Two Pages of the Dresden Codex, a sacred Maya book probably recovered by an early Spanish expedition near Cozumel in the early sixteenth century.

corded below it. This is what we call the "Long Count" system, and after the day and month cycles we've already explored, it completes the triad of elements that make up the ancient Maya calendar.

It was Förstemann, without any knowledge of Maya culture or language, who first decoded the system and realized how it worked. When he started, one important source he had at hand was Bishop Landa's sixteenth-century treatise *Relación de las Cosas de Yucatán*, which he used as a valuable guide for identifying a number of the day and month signs in the Dresden Codex. With that he was quickly able to recognize some Calendar Round dates. He noticed as well how, in a number of pages, Calendar Round records routinely followed the strings of five numbers, each written with bars and dots and never totaling more than nineteen. These, Förstemann reckoned, must be a different sort of calendar system, using a format of place notation to record a quantity of elapsed time from a certain base date far in the past. He also was able to see that some of these same units could be used to write elapsed time between different days in the almanacs of the Dresden Codex and other manuscripts. Gradually he built up a largely mathematical "proof" for a new system, and by 1890 he had worked out all of the basic components of the calendar, now called the Long Count.

At around the same time, across the Atlantic, an American by the name of Joseph T. Goodman was working on many of the same questions that had occupied Förstemann. Like his German counterpart, Goodman did not come from a conventional academic setting; his background was in the newspaper business. He started out in San Francisco before the Civil War and eventually moved to Virginia City, Nevada, where he was owner and editor of the *Territorial Enterprise*. It was there that he gave the young Samuel Clemens his first job as a writer, which established a lifelong friendship between the two. In later life Goodman sold his part of the newspaper and made money

in the local mining business, which allowed him to dabble in archaeology. By the 1880s a number of important publications on the Maya were just coming out, none more important than the works of the British explorer Alfred P. Maudslay, who was documenting the ruins and sculptures of Copan. Maudslay's photographs and

Joseph Goodman, who independently worked out the mechanics of the Long Count.

the meticulous and accurate drawings he published of inscriptions there finally made the study of Maya glyphs possible. Through mutual contacts, Goodman corresponded with Maudslay, and the two men soon collaborated, with Goodman given the task of publishing the first overview of the "archaic Maya inscriptions." Whereas Förstemann had used the Dresden Codex and other manuscripts in his studies, Goodman mostly focused on the texts of Copan, Quiriguá, and Palenque. Working somewhat independently—it's hard to say how much of Förstemann's work he read—Goodman seems to have reached some of the same conclusions about the Long Count, and identified many of the glyphs for the numbers of the Long Count's units, or periods. He also showed that the same system was at work over many centuries of Maya history. Within a few short years, the librarian and the frontier newspaperman had both cracked the basic working of the calendar as recorded in the ancient books and monuments, and thus created the basic framework for nearly all Maya archaeological research ever since. As Goodman described his own contributions to this breakthrough, in his florid newspaper style, "Eureka! The perturbed spirit of the Maya calendar, which has endeavored so long to impart its message to the world, may rest at last."[2] Hyperbole aside, Goodman was right. If there had been a Nobel Prize for archaeology, Förstemann and Goodman should each have gotten one.

Both men realized that the five numbers written in sequence each represent specific quantities of individual time units arranged in a set order, each decreasing in size as one reads along. So, the sequence of bar-and-dot numbers 9, 16, 4, 10, and 8 corresponds to numbered multiples of these five different units. With the help of a handful of historical sources such as Landa, Förstemann was able to determine the exact amounts of each

The five numbers of a Long Count date, from the Dresden Codex.

unit in this place notation system. The last and smallest unit corresponds to an individual day. Above this (if we look at them in reverse order) is a higher unit composed of 20 days, with the next highest a unit of 360 days. Multiplying this by 20, in turn, gives us a period of 7,200 days, and then the highest of the 5 units is multiplied again by 20, each being 144,000 days. The reckoning system makes ample use of base-20 (or vigesimal) calculations instead of the base-10 system we use in our own counting. This all makes considerable sense, given that Mayan and other Mesoamerican languages state numbers using this same base-20 idea.* It turns out that one of the Long Count units should already be familiar to us: the 20-year k'atun, equaling twenty 360-day units, or 7,200 days. This was a standard-length unit of time used throughout Yucatán and by the Itzá Maya for both historical records and prophecies.

The basic building block of the Long Count system is the 360-day period called the tun (based on the word for "stone"). While not the smallest unit of all, the tun nevertheless serves as the conceptual foundation for the whole calendar, with 20 of them making a k'atun; 20 of those, a unit called the bak'tun; and so on. A span of 360 days approximates a solar year only very roughly (that is, it has no basis in astronomical reality), but this certainly was a factor in its being the key unit in the system that, as we shall soon see, is even larger than Förstemann and Goodman ever imagined. The number 360 was also no doubt important to the Maya and their predecessors because of its numerological properties, having an interesting series of prime factors (2^3 x 3^2 x 5), as well as being divisible by many numbers (2, 3, 4, 5, 6, 8, 9, 10, 12, 15, 18, 20, 30, and so on). (It is probably for this reason that 360 also has an important role in our own numerology, as the degrees on a compass, for example.)

So, just how does the Long Count work as a system? That string of numbers written in the Dresden Codex—9, 16, 4, 10, and 8—we typically transcribe as 9.16.4.10.8. They express the multiples of the 5 standard units in this way:

---

* Although Mesoamerican numerology is based on the number 20, the varied languages of the region do not always reflect this. See Yasugi, 1995: 77–105.

9 × 144,000 days	= 1,296,000 days
16 × 7,200 days	= 115,200 days
4 × 360 days	= 1,440 days
10 × 20 days	= 200 days
8 days	= 8 days
Total	= 1,412,848 days

This large tally equals more than 6,707 years. Any Long Count date therefore records *elapsed* time, expressing a number of days counted from a very ancient "zero" date in order to specify an exact day in history or mythology. As we will discuss at more length a bit later, we have a pretty good idea that the "base date" of this Long Count calendar fell on August 11, 3114 BC, a mythological Creation day far earlier than any real event in Maya history. So, when we add 2,449,648 days to this beginning point we arrive at November 4, AD 755, our way of understanding the date recorded as 9.16.4.10.8 in the Dresden Codex. Förstemann, for one, did not advocate any day-to-day correlation between the Maya calendar and our own, but Goodman did, and this came to be refined and confirmed over the following decades. In the wake of their breakthroughs, a number of scholars on both sides of the Atlantic filled in a few gaps, so that by the turn of the last century, the Maya calendar was basically understood exactly as it is today.*

This Long Count calendar operated concurrently with the 260-day and 365-day rounds. However, it was very different in its structure from these two cycles, presenting a more linear reckoning by means of a place notation arrangement that expressed an accumulation of elapsed days from a set starting point. The temporal scope of the Long Count was therefore far, far greater than the 260- and 365-day components of the Calendar Round. The three systems—the Long Count, the 260-day round, and the approximate solar year cycle—together constituted a triumvirate of calendars used throughout Maya history. Any date, whether

---

* Other early works on the calendar, building on Förstemann, include studies by Brinton, 1895, and Goodman, 1897.

it's historical or mythological, could be written using the three calendars together. We transcribe them in the same way, as seen in these four examples, chosen more or less at random:

9.8.9.13.0	8 Ahaw 13 Pop
9.9.2.4.8	5 Lamat 1 Mol
9.12.11.5.18	6 Etz'nab 11 Yax
9.13.0.0.0	8 Ahaw 8 Woh

Notice that these four dates are in chronological sequence. If this seems unclear, look at the second number within each Long Count—the position of the k'atun period—and how they are progressively larger. These four dates are all important points in the history involving arguably the greatest Maya king, a man named K'inich Janab Pakal, who ruled Palenque in the seventh century AD. The first date is that of his birth, corresponding to March 29, AD 603. He took the Palenque throne when a mere twelve years old on the second date, or July 22, 615. He died a ripe old man at eighty years old on August 24, 683 (the third date). Not long after his son assumed kingship, the kingdom celebrated a major station in the calendar, the ending of a k'atun, on March 11, 692 (the final date). Taking into consideration the length of the periods (about twenty years for a k'atun, one year for a tun, and so on), it's not too difficult to do a quick reckoning of how far apart these dates are from another. They cover nearly a century of history at Palenque.

The Beginning Point

My interest in the Maya calendar first blossomed during an early visit to Palenque, the ancient home of the king just mentioned, K'inich Janab Pakal. It was the summer of 1978 and I was twelve years old. My parents had already recognized and encouraged my interest in the ancient Maya, and they had arranged for me to spend the summer working in Palenque with their friend Linda Schele, the noted art historian with a keen interest in all things Maya. One evening in the project house at Palenque,

I sat down with a legal pad, determined to figure out the calendar for myself. (This sounds incredibly nerdy, sure, but there was no Nintendo or Internet to keep me entertained back in those days.) Over the course of ten or so pages, I made my own calendar, writing that day's equivalent in the Calendar Round with its corresponding Long Count, and writing each subsequent day on the lines below. It was a slow, laborious method, and it allowed the varied intermeshed mechanisms of the calendar to take hold in my mind; as you read the calendar specifics, it is worth noting that while it seems complicated, the systems reveal themselves after a little time and consideration.

As a tally of days, the Long Count calendar had some sort of beginning, or base date, after which the days, winals, tuns, k'atuns, and bak'tuns accumulated over time. And as we've already noted, that base date, according to the correlation of calendars first proposed by Goodman over a century ago, corresponds to the day August 11, 3114 BC, two and a half millennia before Maya civilization came into being, and well before complex society of any sort arose anywhere in Mesoamerica. It's a mythical date, which in many of our later discussions I'll refer to as the "Creation base date," to distinguish it from some other foundations of counts that will come into the picture a bit later. Whoever invented the Long Count system clearly opted to have the calendar's *historical* onset take place in midstream, centuries after this base date. Let's say, just for the sake of illustration, that the calendar's inventors lived around 350 BC—an arbitrary choice on my part—and therefore proposed the beginning for their new ritual calendar to fall on an upcoming round date associated with a sacred number, say 7.0.0.0.0. This would have set things off on their regular course, and would also have necessarily set a base date for the entire system, falling exactly 7 bak'tuns into the past, or a little less than 2,800 years. I have no idea why any such certain starting point would have been chosen, but clearly some decision like this had to have been made in Mesoamerica's remote past. The base date of the Long Count system, then, is by nature an artificial construct.

For simplicity's sake, we can write this beginning point as 0.0.0.0.0, as if the Long Count were like a car's odometer. (The Maya wrote it differently, as we'll soon see.) In the juxtaposition of the three calendars, this date falls on the Calendar Round position 4 Ahaw 8 Kumk'u. The day after was 0.0.0.0.1 5 Imix 9 Kumk'u; the next, 0.0.0.0.2 6 Ik' 10 Kumk'u; and so forth. In seventeen days the count reaches 0.0.0.0.19 10 Kawak 2 Pop, with the next being 0.0.0.1.0 11 Ahaw 3 Pop. Notice here that the final k'in number in the Long Count has reverted to zero, and that the Winal, the unit expressing sets of twenty days, now is one. The system accumulates in this way up through the ever-increasing units of the tun, k'atun, and b'aktun.

If one looks closely, it's clear that multiples of 20 constitute this system—20 days make a winal, 20 tuns make a k'atun, and 20 k'atuns make a bak'tun. This shouldn't come as much of a surprise, since 20 is the base unit of most Mesoamerican numerology, not just in calendrical matters. The lone important exception to this pattern in the Long Count is the 18 winals that comprise the 360-day tun. This seems a strange disruption to an otherwise neat and consistent pattern, but it was necessary in order to bring the tun into close approximation of the solar year of 365 days. In a "pure" vigesimal system, the tun place would be a unit of 400 days, the k'atun 8,000 days, the bak'tun 160,000 days. Despite many claims to the contrary, the Maya never used the Long Count format to record quantities of things. It was always reserved for the long-term counting of days.

Just as with the names of the days and the months of the Calendar Round, the words used for the five periods—bak'tun, k'atun, tun, winal, and k'in—were not necessarily the names used by the Classic-era Maya. (They were established in the scholarly literature less than a hundred years ago, some patched together out of odds and ends of colonial Yukatek terminology—in those days, Yukatek was virtually the only Mayan language known to scholars.) Their original names in the classic hieroglyphic texts were probably somewhat different; as best as I understand them, they were *pik(ha'b)*, *winikha'b*, *ha'b*, *winik*, and *k'in*. The original word for the tun unit was probably *ha'b*, which means "year," either as this 360-day period or the more proper 365-day solar year. As was noted a few chapters ago, the hieroglyph for *year* most likely represents a slit-drum, a musical instrument carved out of a log. The reason for this isn't too obvious, but, as I speculated earlier, it may have had to do with the importance of musical performance in the yearly festivals of the *ha'b*, or else it might allude to the passing of the years and seasons as a rhythmic progression, likened to the beat of a ceremonial drum. Some of the higher periods above the tun incorporate the same *ha'b* hieroglyph. The original name for what we now call a k'atun may have been *winikha'b* (twenty years). The bak'tun was a *pik*, a high-order number term that in a purely vigesimal count system means 8,000 (20 x 20 x 20). In the Long Count it had a different sense, marking the third order of the tun count (the tun being the first, the k'atun the second).

First Dates

We tend to view the Long Count calendar as a hallmark of Maya civilization, but there's good evidence that the Maya borrowed the system from their early Mesoamerican neighbors. While we've seen that the origins of the system remain obscure, the first examples of it come from the Isthmus of Tehuantepec, to the west of the Maya region, more or less where the Olmec thrived centuries before the Maya. Did the earlier Olmec invent it? We cannot say, but the geographic distribution of the earliest Long Count dates is suggestive. The calendar was in place and written on stone monuments in the isthmus area by 200 to 100 BC. It may well have existed among the Preclassic Maya at roughly the same time, but we lack good direct evidence of this.

Our earliest instance of a Long Count date comes from a small, unimposing fragment of a stone slab excavated at the site of Chiapa de Corzo, located in the mountains of central Chiapas, Mexico, just outside the modern city of Tuxtla Gutiérrez, the state capital. Clearly this was part of a much longer hieroglyphic inscription now lost. Looking at what remains, we see a vertical sequence of bar-and-dot numbers: first, at the top, a partial number showing two bars (so, 10 or above), followed by 3, 2, and then finally 13. Below this, partially broken, is a record of a day in the 260-day calendar. The Long Count date elaborates on the day sign, so we have a single day recorded in two very different systems: an incomplete Long Count ([?].[?].3.2.13) and a day we can reconstruct

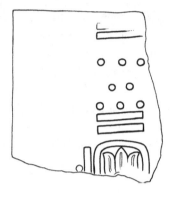

The Chiapa de Corzo fragment, with the earliest known Long Count date. (Drawing by the author)

as 6 Ben. We can identify the day sign because the number of the k'in unit is 13, so it must be the thirteenth of the 20 days, named *Ben* in Yucatán and "Reed" (*Acatl*) among the Aztecs. If the number for the highest period, the bak'tun, is missing, how do we reconstruct it? Well, we can begin by plugging in various likely candidates. Assuming it's a historical date, for example, we can reasonably entertain 7, 8 as possible values for the bak'tun (9 or 10 would be far too late, as indicated by the known style of the carving). The k'atun number is a bit trickier. It's clearly 10 or above, so it could be anything from 10 up to 19. Let's plug in the possible variables and see what we get to choose from:

7.10.3.2.13	5 Ben 11 Kumk'u	8.10.3.2.13	4 Ben 16 Ch'en
7.11.3.2.13	3 Ben 11 Mak	8.11.3.2.13	2 Ben 16 Sotz'
7.12.3.2.13	1 Ben 11 Mol	8.12.3.2.13	13 Ben 1 Kumk'u
7.13.3.2.13	12 Ben 11 Sip	8.13.3.2.13	11 Ben 1 Mak
7.14.3.2.13	10 Ben 16 Pax	8.14.3.2.13	9 Ben 1 Mol
7.15.3.2.13	8 Ben 16 Sak	8.15.3.2.13	7 Ben 1 Sip
7.16.3.2.13	6 Ben 16 Xul	8.16.3.2.13	5 Ben 6 Pax
7.17.3.2.13	4 Ben 16 Pop	8.17.3.2.13	3 Ben 6 Sak
7.18.3.2.13	2 Ben 1 Muwan	8.18.3.2.13	1 Ben 6 Xul
7.19.3.2.13	13 Ben 1 Yax	8.19.3.2.13	12 Ben 6 Pop

Only one date will work—7.16.3.2.13 6 Ben 16 Xul—falling in 37 BC. This is the earliest extant Long Count date so far discovered in Mesoamerica.[*]

[*] The date on the Chiapa de Corzo fragment was first read by Lowe, 1962.

A few other inscribed objects fall very close to the same time. Curiously, none of them is Maya. All are instead inscribed with texts written in another script, called Isthmian, an as-yet-undeciphered writing system used at Chiapa de Corzo and nearby sites from Preclassic to Classic times. Isthmian writing is visually similar to Maya script, most likely recording a different Mesoamerican language family, known as Mixe-Zoquean, perhaps the language of the far more ancient Olmec culture of the Gulf Coast region. For this reason we believe that the Long Count calendar was not originally a Maya invention. The cultural influences from Isthmian and early Maya civilization were intense and long-lasting, including the possible borrowing of the Long Count calendar by the Maya along with other key cultural and religious ideas.*

Once Maya scribes began making use of the Long Count, they opted to insert hieroglyphic terms for each of the individual periods. (No Isthmian texts have corresponding period glyphs.) These glyphs correspond to words for the units of days, largely based on the base-twenty structure for number terms found in many Mesoamerican languages. As we'll soon see, Maya records of Long Counts can assume a very complex and ornate look, usually at the very beginning of a hieroglyphic inscription, where they serve to set the chronological stage for a historical or mythical narrative. Their position at the beginning of texts led early Mayanists to call such Long Count glyphs an "Initial Series."

Long Count dates written in both Isthmian and Maya systems make use of what we call an "introducing glyph," for want of a better term. These appear always in front of (or above) the highest period of the Long Count, the bak'tun, and sometimes they assume a large or overwhelm-

---

* One important Isthmian or epi-Olmec inscription with an early Long Count date is Tres Zapotes Stela C, which is very close in time to the Chiapa de Corzo fragment — only three years later (7.16.6.16.18). This monument was first reported in 1940, when the famous Olmec archaeologist Matthew Stirling proposed its early dating to the first century BC. Many Mayanists thought Stirling crazy to consider such an early date, but he was right. (Discovery of a fragment of Stela C later confirmed his reading.) This gave Mesoamerican scholars the first hint that the Olmec were indeed pre-Maya. See Stirling, 1940.

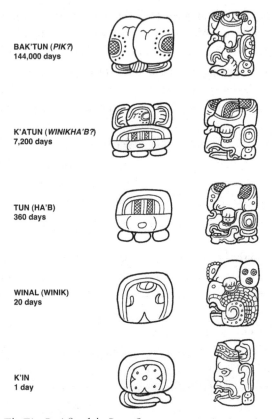

BAK'TUN (*PIK?*)
144,000 days

K'ATUN (*WINIKHA'B?*)
7,200 days

TUN (HA'B)
360 days

WINAL (WINIK)
20 days

K'IN
1 day

The Five Periods of the Long Count. (Drawing by the author)

ing size, a bit like we might find in an initial capital letter in a medieval
or Renaissance manuscript. We have little idea about the direct meaning
of introducing glyphs, and the internal elements are hard to decipher
as word elements. I suspect that the form these glyphs assume in most
Maya inscriptions reflects centuries of visual development, and perhaps
they even lost much of their original meaning as hieroglyphs in this long
process. One interesting feature of introducing glyphs is a variable element at their core, above the sign that resembles a tun glyph. These internal signs show great variation, and in the 1930s the German scholar
Hermann Beyer realized that their forms co-vary with the "month" of the

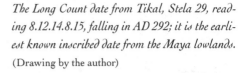

The Long Count date from Tikal, Stela 29, read-
ing 8.12.14.8.15, falling in AD 292; it is the earli-
est known inscribed date from the Maya lowlands.
(Drawing by the author)

365-day calendar. That is, when the date falls in the month *Pop*, the internal element shows a jaguar; when the month is *Yaxk'in*, we see a k'in glyph in the same spot. These variables have come to be called "patrons" of the months, although their exact significance remains unclear. I assume that they are historically connected in some way to Isthmian writing, where the 365-day station of a given date appears inside its version of the introducing glyph, never as a separate block as in Maya. Perhaps if the Maya borrowed features of the Long Count calendar and its format, they retained this original format while also developing their own glyphs for the months, ones reflecting Mayan names and pronunciation. We will come back to the introductory glyph in the next chapter, when we investigate the more expansive records of Long Counts from Coba and other sites.

For now, let's look at some other examples of Long Count or Initial Series dates, to get an idea of the various ways scribes wrote them throughout Maya history. We saw that the scribes of the Dresden Codex wrote them pretty much exactly as we saw at Chiapa de Corzo, as a line of bars and dots, following a format that spanned over a thousand years of history and development. This simple vertical column was in all likelihood the basic format for such dates, established early on in history and also used, one might imagine, in historical and religious record books. The Maya rarely ever used this format in their elaborate textual presentations on stone monuments, however.

A Long Count date painted on a tomb wall at Río Azul, Guatemala, reading 8.19.1.9.13, falling in the year AD 417. (Drawing by the author)

The earliest known Long Count date on a Maya stela comes from the great city of Tikal, in northern Guatemala. It appears on a small stone monument known as Stela 29, discovered by archaeologists in an isolated area of the site, which suggests it was "dumped" in ancient times.[*] On the back of the stone we see a vertical line of glyphs presenting a Long Count date, now showing the heads of animals and fantastic beasts inserted directly after the upright numbers, each standing for (running from top to bottom) the bak'tun (8), k'atun (12), tun (14), winal (8), and, set apart slightly and mostly missing, the k'in (15). Only a small portion of the Calendar Round survives—one dot of the coefficient on the day sign. Because it is the 15th k'in, the day sign must be *Men*. In fact, the complete Long Count and Calendar Round can be easily reconstructed because it is fixed in linear time: 8.12.14.8.15, with the corresponding Calendar Round falling on 13 Men 3 Zip (July 8, AD 292). Unfortunately we are missing the portion of the text that explains what happened on this day, but I suspect it corresponds to the accession of the early Tikal king portrayed on the monument's front. Several monuments from sites not far from Tikal exhibit a style that looks earlier than Stela 29, but none has a date we can pin down.

[*] For a description of Stela 29 and its discovery, see Shook, 1960, and Jones and Satterthwaite, 1980, p. 61.

A Long Count from Pomona, Panel 1, recording the k'atun ending on 9.13.0.0.0, falling in the year AD 692. (Drawing by the author)

A more complete Long Count from a painted tomb at Río Azul, Guatemala, offers a somewhat simpler and standard format for writing a date, using two columns instead of one. The use of two columns, reading across each row and down, is actually far more standard in Maya writing. Here we see different glyphs for each of the five periods, again with bar-and-dot numeral prefixes. Below the large "introducing glyph," the date reads: 8 bak'tuns, 19 k'atuns, 1 tun, 9 winals, and 13 k'ins, or, in standard abbreviated notation, 8.19.1.9.13.

A later example of a Long Count date from the site of Pomona, Mexico, is written as 9.13.0.0.0 8 Ahaw 8 Woh. Here the period signs again are written as heads of various animals and gods—what we

A Long Count date (9.16.15.0.0) written as "full-figure" glyphs.
(Drawing by the author)

call "head variants" on the hieroglyphic forms we encountered at Río
Azul. The numbers written in this date make use of several sequential zeroes, indicating that the date is a period ending, a station of
great significance in the constant accumulation of days the calendar
represents. Any Long Count date that ends in two or more zeroes is
a period ending, meaning that a key station in the continuing tally
of days has been reached. Bak'tun and k'atun endings are especially
important, given their infrequence—bak'tun endings occur only ev-

ery four hundred or so years. Also important were the quarter sub-divisions of the k'atun, that is, where the tun number is five, ten, or fifteen. To illustrate the nature of period endings, it might help to glance over the dates written in Appendix 1 shown on page 317, which lists the significant period ending dates of the Classic period. Nearly all of these are recorded in one or more inscriptions from various ancient sites.

In a few instances, ambitious scribes went to extremes in terms of artistry and visual complexity to write Long Count dates. We see this most clearly on the stelae at Quiriguá and Copan, in the south-eastern Maya area, where both numbers and periods are gorgeously written with "full figure" elements. The glyphs are not just heads, but complete interactive bodies, usually combining two or more separate elements—say, a number and a period glyph. For example, one record of the Long Count date 9.16.15.0.0, written on Stela D from Quiriguá, assumes the form of five impossibly complex grid blocks, within which we see writhing and gesturing figures. By looking closely at the di-agnostic features of each figural sign, we recognize the numbers and periods themselves. Starting in the second block, after the introduc-ing glyph, we have: 9 bak'tuns, 16 k'atuns, 10 tuns, 0 winals, and 0 k'ins. In substance this is no different from a string of bars and dots as Förstemann saw in the Dresden Codex. Few artisans went to such elaborate lengths, however, and in their day, the scribe and carvers of Stela D were unmatched in skill and design sense. Arguably, these and other full-figure Maya glyphs at Quiriguá and elsewhere represent the most visually intricate written forms ever devised in human history.

Linking the Long Count and the Calendar Round

The Long Count and Calendar Round are numerically linked in certain fundamental ways. As we've seen, any Long Count has a corresponding Calendar Round to which it is inalienably tied, and both systems move forward in tandem as each day goes by. Again let's look at a sequence of days from Classic-period history to see just how they move along together. The first of the dates that follow is the birth date of a famous queen from the site of Piedras Negras; the sequence thus represents the first week of her life.

9.12.2.0.16	5 Kib 14 Yaxk'in	July 2, 674
9.12.2.0.17	6 Kaban 15 Yaxk'in	July 3, 674
9.12.2.0.18	7 Etz'nab 16 Yaxk'in	July 4, 674
9.12.2.0.19	8 Kawak 17 Yaxk'in	July 5, 674
9.12.2.1.0	9 Ahaw 18 Yaxk'in	July 6, 674
9.12.2.1.1	10 Imix 19 Yaxk'in	July 7, 674
9.12.2.1.2	11 Ik' Seating of Mol	July 8, 674
9.12.2.1.3	12 Ak'bal 1 Mol	July 9, 674

Notice first how each Long Count has its corresponding Calendar Round, and how they each move forward by one day. Also notice how the day names are all keyed directly to the number presented in the k'in position of the Long Count, with 1 signaling *Imix*, 2 the day *Ik'*, 0 the day *Ahaw*, and so on.

An interval of 18,980 days, or about 52 years, falls between any two sequential Long Count dates that share the same Calendar Round. To pick an arbitrary example:

9.13.8.15.3 6 Ak'bal 11 Muwan
9.16.1.10.3 6 Ak'bal 11 Muwan
Interval of 2.12.13.0 = 52 x 365 days = 18,980 days

The figure 18,980 is a product of three factors: the days of the vague solar year (365), the number of day signs that can occur with a given position in the solar year (4), and the possible coefficients of those days (13). This same 52-year cycle of the full Calendar Round was the basis for the Aztec "New Fire" ceremony, when the world was symbolically cleansed and renewed on the anniversary of the year named 2 Reed, when the heavens were created and the god Tez-catlipoca was born.

Historical anniversaries based on the units of the Long Count were always of great interest to ancient Maya historians and chroniclers. Many records found on stone monuments at sites such as Tikal, Yaxchilan, and Piedras Negras make many references to days when a king was, for example, "two k'atuns in the rulership," or when "his three k'atuns ended" (meaning three k'atuns of life, from his birthday). As we'll examine more closely in chapter 9, this constant framing of royal life events in terms of the calendar and its structure probably reflects an intimate conceptual link in ancient Maya thought between the person of the king, the *ajaw*, and the time periods themselves.

A lengthy inscription from the Classic era typically contains many dates, especially if it's a complex historical or mythological narrative. An opening "Initial Series" Long Count might appear at the very beginning, and then, after a while, reading down usually in double columns, we come across numerous day and month signs of the Calendar Round, usually recognizable by the distinctive cartouche around the day sign (see figure showing the Dos Pilas monument on page 58). But a string of dates is not enough. A good Maya scribe would also want to take great care to express the precise intervals between these dates, and to do so, he used the very same units of the Long Count. An interval between two important days in a historical narrative might not be very long at all, so the units employed might be only, say, the tun, winal, and k'in. Some of these intervals, called "distance numbers," can be only a few days or even one day. It all depends on the nature of the story being told. When a distance number is used, it's typical to have it written in reverse order: k'in, winal, tun, k'atun, etc. This is a bit confusing at first for the new student of Maya writing and numerology, but one gets used to it, and can start doing calculations when confronted with a number of different Calendar Rounds. Distance numbers come in very handy for ancient and modern readers alike, for they set the narrative very clearly in time, anchored to the opening Long Counts of the text. Then if one comes across a Calendar Round later on, it's clear where it exists in time, so many tuns and k'atuns after or before the date written before it.

Period Endings

As mentioned earlier, the most significant stations of the Long Count are what we call period endings, when periods such as the tun, k'atun, or bak'tun reached a rounded position, usually celebrated most when their number was divisible by five. For example, the Long Count date 9.16.5.0.0 8 Ahaw 8 Sotz' is a period ending, when the first five tuns of the k'atun were completed. Other period dates are easy to spot, such as 9.6.0.0.0, 8.19.10.0.0, or 9.12.0.0.0. The Maya certainly took notice of these subdivisions of time, each of which was in essence an anniversary of the "zero day" of Creation, worthy of numerous ceremonial events and performances. The turn of the k'atun predicted by the Itzá prophecies is a good example of a period ending in more historical time.

Not all period endings were created equal. Tun and k'atun endings falling on 0, 5, 10, 13, and 15 seem to have been especially important. Perhaps the most ritually significant k'atun ending of the Late Classic period was the end of k'atun 13 on 9.13.0.0.0 8 Ahaw 8 Woh, which fell on March 13, AD 692. This was celebrated throughout much of the Maya world as an important "cosmic" number, for as we have seen, the number 13 holds obvious importance in archaic Maya numerology, with its associations with the date of Creation (13.0.0.0.0), as well being a main factor in defining the 260-day sacred round. With the approach of 9.13.0.0.0 8 Ahaw 8 Woh, a number of kingdoms prepared great sculptural and architectural monuments. At Palenque, for example, the local king and his priests and nobles constructed three elegant temples for the occasion, each dedicated to one of the city's patron gods known as the Palenque Triad (we will revisit them in chapter 7). The three pyramids are today collectively known as the Group of the

Cross, and the carved tablets in each recount an epic myth concerning the origins of the gods in primordial time. In some real sense, I believe that the Maya of Palenque viewed the historical date 9.13.0.0.0 as a numerological reflection of that distant Creation, which motivated the construction of what amounted to three cosmic temples associated with the origin of the world.

Each and every period-ending date, such as 8.19.0.0.0 or 9.16.10.0.0, falls on the day Ahaw in the 260-day cycle. The reason for this is purely mathematical, since both the winal period and the list of day names span twenty days. This juxtaposition of number and name created an extremely important symbolic association where the day meaning "Lord" would forever be linked to the key station of the Long Count (*Ajaw* being the ancient variant of the later day name *Ahaw*). In this way, the k'atun periods and their subdivisions were "ruled" by particular Ahaw dates.

As the prophecies of Yucatán and the Itzá showed us, a cycle of thirteen k'atuns was of particular importance in Yucatán in the years before and after the Spanish Conquest. Why thirteen? This comes from the coefficients of the Ahaw days, which of course can only be from 1 through 13. Thus, after 13 k'atuns pass (roughly 260 years), time arrives at a new period with precisely the same Ahaw number, a "repeat" of history that formed the basis of the fusion of history and prophecy in later Yucatán, and perhaps among the classic Maya as well. We know of no name for this period, although some now prefer to call it a *may* cycle, but based on little evidence.*

---

* The so-called *may* cycle is a misnomer. It was first proposed by Munro Edmonson, 1979, 1982, based on faulty readings and overextended arguments about the history of the Mesoamerican calendar. In highland Mayan languages, the word *may* serves as a term for "twenty (years)" and therefore seems to have been equivalent to the k'atun of ancient Yucatán. The erroneous concept of a *may* cycle was most recently adopted by Prudence Rice in her elaborate overview of ancient Mayan political structure. See Rice, 2004, 2008. The cycle of thirteen k'atuns was important because of its numerological character, but in ancient inscriptions we see little if any evidence that it ever formed the basis of geopolitical structure, as it seems to have done in the fifteenth and sixteenth centuries.

Period endings in the Long Count were the greatest of ritual occasions for Classic-era Maya kings. Nearly all of the stone stelae at sites such as Copan, Tikal, and Yaxchilan were meant to commemorate these days and, most especially, the ceremonies that the rulers oversaw in their celebration: casting incense, drilling fire, sacrificing war captives, as well as in a rite called the "binding of stones." As we will see when we look at these rituals in more detail in chapter 9, one of the principal duties of Maya kings—in addition to ruling the community, waging war, and the like—was to tend to time, ensuring its good health as yet another manifestation of *k'uh*, the sacred order of things.

Finding a Match

In the wake of the great breakthrough by Förstemann in clarifying the workings of the calendar, a basic question came naturally to the minds of many scholars: How can one arrive at a direct correlation of a Maya date with one in our own calendar? Many different proposals have been put forward over the last century or so, but it wasn't until midcentury or later that Mayanists reached widespread accord about the notorious "Correlation Question," although the issue is still debated to this day. The story of how scholars proposed various solutions and eventually reached a generally accepted answer is hardly dull or overly mathematic, and is full of academic politics and colorful characters.[*]

Once the Long Count and the Calendar Round had been unlocked and studied, archaeologists quickly saw how useful written dates could be in anchoring a relative chronology among Maya monuments and even between Maya sites, even if the correlation remained vague and unclear. Assuming that ancient scribes at distant sites were using the same calendar—a fairly safe assumption, and long since confirmed—early scholars sought out all of the available drawings and photographs of inscriptions to look for dates, with the simple aim of

---

[*] For general discussions of the correlation question, see Thompson, 1935, 1950; Kelley, 1983; and Lounsbury, 1978, 1992.

developing a bigger picture of what came before and after what. In this way, even if no one was sure of the right correlation, they could be secure in knowing that the Long Count 9.15.0.0.0 came after 9.13.10.0.0, and that both in turn fell many years after 9.8.0.0.0.

It was Goodman who developed a correlation between the calendars that still stands as the most reasonable and widely accepted. He did so through a logical sequence of inferences based on clues he saw in historical documents written by the Maya of Yucatán in the early colonial period. Some of these documents recorded the arrivals of Spanish conquerors in terms of native chronology, stating, for example, that the founding of the capital at Mérida was in the year beginning 13 K'an—that is, 13 K'an 1 Pop. (These and other dates listed here use the Yukatek system of dating, where the month position is one day off the standard classic calendar.) In European history this event is known to have taken place in 1541. Furthermore, Goodman noticed that documents relating to the history of the noble Xiu family mentioned that an important nobleman named Ahpula died on the day 8 Imix 18 Sip within the year beginning with 4 K'an 1 Pop, which started on July 16, 1545. This meant that Ahpula perished on September 11, 1545. Cross-referencing this with other records, Goodman noted also that the chronicles gave Ahpula's death as occurring in the sixth year after a k'atun ending on 13 Ahaw. Just based on the math, extrapolating backward, this could only mean a placement on 13 Ahaw 7 Xul, which fell on October 30, 1539. The task at this point was a simple one: finding a k'atun ending in the span of Maya time that would reasonably fall on this very Calendar Round date, or what the Classic-era Maya would have written as 13 Ahaw 8 Xul. Goodman found this:

> 11.16.0.0.0 13 Ahaw 8 Xul   October 30, 1539

Now he had something. The majority of Long Count dates written on stelae at archaeological sites were in bak'tun 9—dates such as 9.16.0.0.0, for example. Finally there was an absolute anchor to use in calculating backward some eight centuries to reach the Classic era. As Goodman succinctly put it in his brief paper of 1905, "The result shows that Copan, Quiriguá, Tikal, Menche [Yaxchilan], Piedras Negras, and the other more modern capitals, flourished from the sixth to the ninth century of our era, speaking in round terms." Goodman had outlined his basic argument for a correlation, but not everyone was yet convinced that his was the right answer. There were enough inconsistencies in other dates recorded in the Xiu family chronicles to call for caution.

Mayanists who have long debated the correlation issue routinely refer to something called a Julian day number, often also used by astronomers as a standardized, number-based calendar for calculations and observations. Simply put, the Julian day number represents the number of days elapsed from a base date on January 1, 4713 BC. With this number, any correlation of Maya and Christian calendars could simply be referenced by another number, a "correlation constant," that would bridge a given Long Count day number with the Julian day number. Although he didn't use it, Goodman's proposal involved a correlation constant of 584,280 days. When this is added to the Long Count day number, we get the Julian day number, and thus a link between the two systems.

When Goodman made his correlation proposal, two Harvard students, Herbert Spinden and Sylvanus Morley, took great interest and would spend much of their later careers pondering and testing the correlation idea in different ways. Spinden had a fascinating back story before ending up at Harvard. He was born on the South Dakota frontier in 1879—just about when Förstemann began his studies of the Dresden Codex—and his early life there created a deep interest in the arts and ways of the Native Americans, not to mention a strong sense of adventure. In 1900, Spinden left his job on the Dakota railroad to participate in the Alaska gold rush, but he returned a year or so later, empty-handed. Just how he ended up at Harvard in 1902 isn't very clear, but his love of Native American culture steered his studies there in anthropology and archaeology.

A couple of years into his Harvard studies, Spinden met "Vay" Morley, an enthusiastic and tireless classmate a few years younger; the two became longtime friends and colleagues in Maya studies. Morley came from an academic family, and also had spent a good part of his youth in the West, in Colorado. Each traveled far from Cambridge, Massachusetts, in summer to participate in excavations of Pueblo ruins in the American Southwest, and soon they each journeyed farther south, into Mexico and Yucatán. Eventually, too, during World War I, Spinden and Morley worked together as U.S. intelligence agents, collecting information on possible U-boat hideouts on the Mexican coast for the U.S. Office of Naval Intelligence.*

In those early years of field archaeology, not much was known about the prehistory of Native Americans. Just how old were the Pueblo sites in New Mexico? And how old were the magnificent Maya ruins in Yucatán? A hundred years ago, basic chronology was *the* overarching concern of archaeological field research, and Spinden and Morley, among others, always veered far from their intense interest in the subject. Spinden's dissertation at Harvard, *A Study of Maya Art,* was a wonderfully sophisticated work that's still of great use, and which I consult from time to time in my own research. In it, he uses Long Count dates on monuments to establish the evolution of Maya art styles, and therefore extend their usefulness as a gauge for the relative dating of sculptures and architecture. Morley, for his part, went on to build a great career as a field archaeologist and explorer. After the war, he initiated an ambitious excavation project at Chichen Itzá, and for years he engaged in a tireless quest to explore the most remote areas of the Maya jungle, looking for inscriptions there—"to bring home the epigraphic bacon," as he once wrote. Morley's bacon was documenting Long Count dates; he had little interest in the "non-calendrical" portions of the inscriptions. It became a near obsession for the intrepid, energetic scholar.

Morley wasn't convinced of Goodman's correlation, thinking it placed some Maya dates uncomfortably late in his vision of Maya history. In 1909 he proposed an alternative, but this he later rejected, expressing frustration that a reasonable argument for a correlation would never be possible. To him, the dates in the colonial sources were simply a mess—and he was right to a great extent. Nevertheless, Morley's general doubts about Goodman seem to have inspired Spinden,

---

* See Harris and Sadler, 2003, for a fascinating account of Morley and Spinden's adventures as intelligence agents.

who at war's end took up the task of coming up with an alternative to Goodman's proposal. In 1924 he wrote a wide-ranging essay in support of this earlier time frame, advancing and refining a correlation that positioned dates 13 k'atuns (256 years) earlier than in Goodman's scheme. (Spinden's correlation constant number was 489,384.) Such a time frame could accommodate the important statement in the Xiu histories that a k'atun ended on a 13 Ahaw, near the death of Ahpula.

For a short time in the late 1920s, Spinden's correlation—sometimes called the Morley-Spinden correlation—gained some acceptance among archaeologists, but the honeymoon didn't last long. Shortly after he published his thesis, rejoinders came on two fronts. In Mexico the scholar Juan Martínez Hernández published a new analysis of documentary sources in 1926, reaffirming Goodman's correlation, but with a slight adjustment of one day. And shortly thereafter, the young J. Eric S. Thompson—soon to be among the top Mayanists of the twentieth century—offered his own rejection of Spinden's position, generally agreeing with Goodman and Martínez Hernández, suggesting his correlation standard of 584,285 (five days off Goodman's original).[3]

The fatal blow to Spinden's calendar correlation came in 1930, when John Edgar Teeple published his keen analysis of Maya moon-age records.[4] A great many Long Count dates are accompanied by statements of the age of the moon, always recorded as a number from one to twenty-nine. Teeple proved to everyone's satisfaction that these stand for elapsed days within a lunar month of twenty-nine or thirty days, counted (as we now know) from the first appearance of the new moon. Teeple's discovery provided an important check for any proposed correlation of the calendars, and Spinden's proposal failed completely. His correlation was widely rejected in favor of Goodman's older idea, which, with a little tweaking, did conform nicely to Teeple's lunar data. In this way, the "Goodman-Martínez-Thompson" correlation (GMT as insiders call it), or some close variation of it, gained wide acceptance. In the face of mounting counter-evidence, Spinden stubbornly held fast to his correlation and never accepted that it was off base. Dejected and perhaps even alienated, he gradually moved away from Maya studies.

Some years later Thompson did offer one slight modification to his original correlation proposal, suggesting a two-day shift from the initial idea in order to bring it in line with the 260-day calendar just discovered in the 1920s and '30s as still in use in the highlands of Guatemala. This "modified GMT," as it came to be called, has a correlation constant of 584,283, and this is widely used by Mayanists to this day.

One compelling piece of evidence backing the general GMT correlation was a date recorded on an inscription from a ruin named Santa Elena Poco Uinic, located in the remote mountains of highland Chiapas. The site is still very difficult to get to, and few archaeologists have ever visited the place. Photographs and drawings of the fragmented monument were published, and Teeple was drawn to one written date accompanied by a glyph showing a k'in "sun" sign surrounded by two elements draped over its top and sides. The date was readable as 9.17.19.13.16 5 Kib 14 Ch'en, which in the modified GMT correlation fell on July 13, AD 790. Thinking that the "covered sun" might well refer to a solar eclipse, Teeple checked the dates and found that, in fact, a total solar eclipse passed over southern Chiapas on July 16. The three-day shift may be more fitting with the original GMT (584,285), but it's close and, I think, an intriguing piece of the puzzle in support for the general veracity of the GMT proposals. Much later, in the 1980s, Yale anthropologist Floyd Lounsbury pointed out that the original GMT also made a better fit with his analysis of the so-called Venus tables of the Dresden Codex.[5] He pointed out that the important date, 10.5.6.4.0 1 Ahaw 18 K'ayab (November 20, 934), corresponded to a heliacal rising of Venus in AD 954, although others have pointed out that the modified GMT can explain the Venus table in other, even perhaps better ways.[6]

There's a very high likelihood that one of the two variants of the GMT correlation (584,283 or 584,285) is the correct one. I'm sometimes asked which one I prefer, but it always seems to me a somewhat odd question, based more on our own Western desire for scientific precision than on any consideration of Maya time and its cultural roles. The modified GMT (584,283) has its strong advocates, based largely on astronomical arguments, and it has also gained favor among those who study the modern calendar in highland Guatemala, and among some historians who study early colonial documents. The real difference may turn out to be that *both* are correct, and that we need not assume a perfect, unbroken, and consistent usage throughout Mesoamerican history. One solution may be preferable for studying the Classic Maya, whereas the other reflects a slightly different calendar in use in central Mexico and highland Guatemala (where Mexican influence was profound in the Late Postclassic period). If I were to choose one over the other—and inevitably I'm in such a position, writing this book—I'd opt for the modified GMT (584,283), since it's in sync with the day count discovered by La Farge, Lincoln, and others, and is still in use today. My sense, though, is that a two-day difference doesn't matter too much, if it can ever be resolved. As my colleague Anthony Aveni well put it, "Astronomical data alone fail to resolve the two day discrepancy between viable correlation constants given that no naked-eye astronomical events are predictable to an accuracy capable of differentiating between the two."[7]

Having said all that, I ought to mention that there's always been a small sliver of doubt in my mind about the GMT. We might tout one of its versions as *the* correlation, but in fact it remains nothing more than the best educated guess at the present time. None of the documentary, astronomical, or other scientific evidence for any correlation proposal, including the GMT, is completely airtight, and it is best that we keep an open mind for now, making use of it, always questioning it, and entertaining any reasonable alternatives that may come down the line. And as readers of this chapter can certainly attest, the whole messy issue can be mind-numbingly dull, even if important to how we understand ancient Mesoamerican history and culture. I suspect that niggling aspects of the "correlation question" will dog Maya studies for some time to come.

We've encountered dimensions of Maya time that on their face look inherently cyclical (the Calendar Round) and linear (the day tally of the Long Count), but in fact it's difficult to distinguish between the two concepts. The Long Count is a mechanistic, linear-looking progression of years and days almost too abstract to be appreciated in its full scale, yet its building blocks are smaller units that operate cyclically. Our numeration system is somewhat similar in expressing "linear" accumulations of numbers, but each place unit within a large number involves constant, cyclical repitions of the individual numbers. In this way, as we have seen, numbered k'atuns, winals, and so on can recur many times if one looks at time on a large enough scale. Perspective is key in being able to discern whether time operates in a linear or cyclical way; up close it may look linear, but cycles appear and reappear as the larger system comes into proper focus. For example, a thirteen-tun station will occur every two decades, and two, three, or four others could be experienced in anyone's lifetime. A bak'tun is not experienced in the same way: an ancient Maya daykeeper who lived in the era of 9 bak'tuns would never live to experience its changeover to 13 bak'tuns. Even if one could live for centuries, one would probably perceive such cycles as being as fairly linear, given that a cycle of 13 bak'tuns takes almost 5,129 years to complete. One perceives such time scales as moving along in one way, given that its turnover means little to human experience.

There's an unfortunate tendency to see "cyclical" or "linear" cognitive models of time as two exclusive types. So-called cyclical concepts of time were widespread in ancient cultures such as those in early India and Greece, and were founded on the idea that time's passage was inherently repetitive, both in the long and short term. Just as the passing days present an obvious cyclical structure, so, too, do larger periods, epochs, and even events in history. Linear time is often placed in utter contrast to cyclical time, and is said to have its roots in Judeo-Christian philosophy and concepts of history, where time has a single beginning and subsequent end, with a one-way progression in between. Not surprisingly perhaps, many historians and philosophers therefore invoke linear time as inherently more rational and scientific in its outlook, an important foundation of modernist thought and of notions of universal and human progress.[*]

To me it's all a red herring. A key difference between linear and cyclical comes down to whether one considers the progress of time to be a simple one-way street or as capable of accommodating more complex notions of multiple "beginnings." Any discussion of these concepts

[*] See Gould, 1988, for a wide-ranging look at this issue, especially as it relates to the study of geologic deep time.

within our own science and philosophy is beyond the scope of this book, not to mention my own expertise, but I do find it fascinating that modern theoretical physics and cosmology now entertain "cyclical" models of the universe that have little in common with traditional accounts of how the universe began and evolved. In the conventional model, the "singularity" of the Big Bang stands as time's universal starting point. But modern science also debates and contemplates more radical notions of "multiverses," of various universes existing in sequence or concurrently with our own. No longer just the realm of science fiction, such thinking, even if it turns out to be wrongheaded, has entered our highest, most cutting-edge philosophical considerations of the universe.

An ancient Maya *aj k'in* priest would no doubt look on this with interest and approval.

7

BEGINNINGS AND ENDINGS
OF THE WORLD

*The world was not lighted; there was neither day nor light nor moon.
Then they perceived that the world was being created. Then creation
dawned upon the world. During the creation thirteen infinite series
added to seven was the count of the creation of the world. Then a new
world dawned for them.*

—*The Book of Chilam Balam of Chumayel*[1]

The Mesoamerican worldview saw the universe as an evolving and developing system, punctuated by episodes of destruction and reformulation as it lurched toward the historical present. Narratives from the Maya, the Aztecs, and other traditions tell how, over vast time periods, the gods performed multiple experiments with the creation of people until they got it right. The quasi-people who live in trees, howler monkeys and spider monkeys, were the leftovers of one such trial effort, when the gods' attempt at making people somehow fell short. Finally, after wiping the slate clean a few times, they honed their skills and formed humans who could speak, reason, and, most important, perform the proper veneration of the gods. From these stories, still found in many parts of Mexico and Guatemala, we get a strong sense that the purpose of newly minted humans was to worship and sustain the gods as part of an overarching contract, a covenant with the forces of nature who would sustain them in return.

I find it interesting that in very general terms, Mesoamerican myths describe the process of "Creation" as a series of starts, destructions, and restarts, a track of development that's ever improving from early tree-bound quasi-people to the humanity of the present world. It sounds not unlike our current understanding of evolution's "progress" and deep-time history. Our science informs us that the earth's history has been an often violent story of natural experimentation and selection, with life formed, made extinct, and re-formed on multiple occasions. Maya and Aztec mythology is no anticipation of Darwinian philosophy, of course, but I think such stories reveal how human minds across time and space are capable of making insights into the processes of the world, even if they represent them and retell them in very different ways.

This recurrence of creation and destruction fits well with the idea that the present world may not have a well-defined ending. That is, most mythical cycles of past world beginnings and endings are meant to explain how people and the order of things got where they are *today*. They don't necessarily point at all to what will happen when our world comes to an eventual end. Mesoamerican accounts of world origins are about primordial time, sketching out for us like a preamble a just-so story about what led up to the current condition of the world.

Because different Mesoamerican cultures produced different accounts of how the world came to be, we'll treat them separately, beginning with the myths of the Aztecs and then seeing how they compare with those of the Maya.

Aztec Accounts of Creation

The rich assortment of documentary sources that survive concerning Aztec history, culture, and religion includes one remarkable vision of world Creation quite different from the myths we know from the Maya. According to the Legend of the Suns (*Leyenda de los Soles*), the historical Nahua and Mexica people lived in the age of the Fifth Sun, the last of five distinct eras that came to be created and destroyed.

The best account of this epic story comes from a Nahuatl manuscript written in 1558, translated here:

Here is the account of the learned ones. A great time ago he [the Sun] formed the animals and began to give them food to eat. Thus it is known that this same Sun gave beginnings to so many things, 2513 years ago, prior to this day, the 22 of May, 1558.

This Sun named 4 Jaguar lasted 676 years. Those who died then for the first time, in this age, were devoured by jaguars in the Sun of 4 Jaguar. And they ate 7 Grass, which was their nourishment, and they were devoured after thirteen years; all perished and was ended. Then the Sun disappeared. It was the year was 1 Reed. They began to be devoured on the day with the sign 4 Jaguar. In this way it ended and all perished.

This sun [is] 4 Wind. Those who died a second time were carried away by the wind, in the Sun of 4 Wind. As they were carried away they were turned into monkeys, and their houses and trees were all blown away by the wind. And they ate 12 Snake as their nourishment. They lived for 364 years until the one day when they were carried off by the wind, perishing on the day 4 Wind. Its year was 1 Flint.

This Sun is 4 Rain. And those who lived in the Sun of 4 Rain, the third [Sun], were destroyed because it rained fire upon them and they were turned into turkeys. The Sun burned, as did their houses. They had lived for 312 years, until that day when it rained fire. They ate 7 Flint as their nourishment. The year was 1 Flint, on the day 4 Rain, when those who were turkeys (*pipiltin*) perished.

The name of this Sun is 4 Water, because there was water for 52 years. Those who lived in the fourth Sun, 4 Water, lasted for 676 years until they were destroyed. And they perished by being overwhelmed by the flooding water, when they were transformed into fish. The heavens came down upon them on that day (4 Rain). And 4 Flower was what they ate as their nourishment. The day was in the year 1 House, on the day 4 Water, when they all perished, when all the mountains disappeared, because it flooded water for 52 years.*

The name of this Sun is 4 Earthquake, the one in which we now live. And here is its sign, of how the Sun fell into the fire, into the divine hearth, there at Teotihuacan. It was also the Sun of our Lord Quetzalcoatl in Tula. The fifth Sun, 4 Earthquake, is called the Sun of movement (*ollin*) because it moves and follows its path. And as the elders continue to say, under this Sun there will be earthquakes and hunger, and then our end shall come.

Some of the language in this account ("the name of *this* Sun . . .") offers a subtle clue that the surviving text might well be a transcribed reading or description of a pictorial document, not unlike one of the screen-fold manuscripts or codices. Other surviving accounts of the same mythic story offer much of the same essential information as we read in this version, although sometimes a few interesting details appear in some sources that are omitted in others. For example, we read in an alternative version that each of the "suns" was overseen or embodied by a different Aztec god.

The curious names of each ancient time period—4 Jaguar, 4 Wind, 4 Rain, 4 Water, and 4 Earthquake (or Movement)—clearly make reference to the manner of their respective destructions. Each age had a preordained death sentence of sorts, symbolized by its divine, time-based label. These names correspond of course to calendar terms in the sacred divinatory cycle of 260 days. Their repeating number four prefixes also emphasize the quadripartite symmetry that was so basic to Mesoamerican cosmology and thought.

The time spans of these previous four world eras are not particularly

---

* The English translation of the *Leyenda de los Soles* is my own, based on the Spanish and English versions of Velázquez, 1975, and León-Portilla, 1974, respectively. There are several ambiguities in the original Nahuatl, which have led to problems in each of these translations. Here I have attempted to combine the positive aspects of each work.

long (676, 364, 312, and 52 years, respectively), suggesting to me that they may be somehow rooted in historical reality, a rough conceptualization, perhaps, of the durations of pre-Aztec cultures who lived in Mesoamerica. The Aztecs were keenly aware of the earlier civilized "Toltecs" and of the vast ruins that surrounded them on the landscape, such as Teotihuacan and Tula. Perhaps their myths of previous creations and destructions helped to explain the abandoned cities and buried remains they everywhere encountered. Aside from such speculations, the lengths of the four previous ages are mathematically and cosmologically meaningful, reflecting the importance of another number in Mesoamerican thought: 13. Each of their durations is shorter as time progresses, and each is evenly divisible by 13: 676 (13 x 62); 364 (13 x 28); 312 (13 x 24); 52 (13 x 4). The final age, destroyed by floods, lasted a mere 52 years, a familiar span corresponding to the Aztec "century," formed through the convergence of the 260- and 365-day cycles and celebrated by the all-important New Fire ceremony, a time of world creation and rebirth.

The Legend of the Suns and other Aztec sources state that the fifth and last "sun" was called 4 Earthquake, or *Nahui Ollin,* in Nahuatl ("our time now, in which we now live. It is the sign that fell into the fire of the Sun, in the divine hearth at Teotihuacan."). This calendar sign was not only a designation or name for the present era of humanity, but also the name applied to the resplendent sun in the sky, personified and embodied as the light-haired and bejeweled deity Tonatiuh, "the soaring eagle, the turquoise prince." As one account notes:

> Every two hundred and sixty days, then (the Sun's) festival was honored and celebrated. They observed it on his day sign, called Nahui Ollin. And before his feast day had come, first, for four days, all fasted. And when it was already his feast day, when first he came forth, when he emerged and appeared, incense was offered and burned; blood (from the ears) was offered. This was done four times during the day—when it was dawn; and at noon; and past midday, when already (the sun) hung (low); and

when he entered (his house)—when he set; when he entered (his course).*

The mention of Teotihuacan as the place where the sun "fell into the divine hearth" clearly marks the fundamental importance of that ancient city in Aztec thought. According to another valuable Aztec account recorded by Friar Bernardino de Sahagún, this is where the gods convened to create a new age, after the destruction of the previous four:

> ... at this place (the divine hearth) the fire blazed for four days ... then the gods spoke. They said to Tecuciztecatl, "Now, Tecuciztecatl, enter the fire!" Then he prepared to throw himself into the enormous fire. He felt the great heat and was afraid. Being afraid, he dared not hurl himself in, but turned back instead. ... Four times he tried, four times he failed. After these failures the gods spoke to Nanahuatzin, and they said to Nanahuatzin, "You, Nanahuatzin, you try!" And as the gods had spoken, he braced himself, closed his eyes, stepped forward, and hurled himself into the fire. The sound of roasting was heard, his body crackled noisily. Seeing him burn thus in the blazing fire, Tecuciztecatl also leaped into the fire. When both of them had been consumed by this great fire, the gods sat down to await the reappearance of Nanahuatzin; where, they wondered, would he appear? Their waiting was long. Suddenly the sky turned red; everywhere the light of dawn appeared. It is said that the gods then knelt to await the rising of Nanahuatzin as the Sun. All about them they looked, but they were unable to guess where he would appear. Some thought that he would appear from the north; they stood, they looked to the north. Towards noon, others felt he might emerge from anywhere; for all of them, everywhere was the splendor of dawn. Others looked toward the east, convinced that from there he would rise, and from the east he did. It is said that Quetzalcoatl (Feather-Serpent), also known as Ehecatl (Wind), and another god called Totec, had guessed the place from which

* Sahagún, Florentine Codex, Book VII, p. 1.

he would rise. . . . When it appeared, it was flaming red; it faltered from side to side. No one was able to look at it; its light was brilliant and blinding; it rays were magnificently diffused in all directions.*

Thus the minor god Nanahuatzin was transformed into Tonatiuh, the Sun who gives light and life in the current world. Sacrifice is a key theme of this story, of course, and through this primordial myth the Aztec state religion drew some of its theological inspiration, surely using the story of world creation as a template for the frequent human sacrifices—all renewal rites in their own right—conducted in the many temples of Tenochtitlan.

None of the Aztec sources say just when the age of 4 Earthquake will end, if it hasn't happened yet. The obvious pattern of predetermination, of naming an era by a day name signaling its mode of destruction, should nevertheless give us some idea how this last creation should someday meet its end.

Nowadays, some pseudo-scholars take the end of the Legend of the Suns as a harbinger of the doom that awaits us in 2012. But such claims are grossly off base for the simple reason that the year 2012 relates to *Maya* accounts of cosmology, and has nothing at all to do with what the Aztecs might have predicted for the future. In reality, the two mythic systems have little connection with one another, and the extant Maya sources never mention anything about an age of "4 Earthquake" or "4 Water" or what have you. This hasn't prevented some New Age writers and Hollywood producers from simply melding these traditions into one pan-Mesoamerican hodgepodge, taking the specificity of the 2012 date from the Maya and grafting it onto the dramatic Aztec imagery of world destruction. There are many problems with such loose distortions of the evidence, of course, not the least being that the upcoming turn of the bak'tun period in 2012 was never described by the Maya as a destruction of anything. But more on that later.

Despite its importance in Aztec cosmology, the Legend of the Suns isn't depicted in any codex or pictorial document; it survives only in

* This translation of Sahagún's account of the myth is adapted from León-Portilla, 1974, pp. 44-45.

Page 30 of the Borgia Codex, possibly depicting the mythological birth of the day count.

written form, transcribed decades after the Conquest. But we do have imagery of other important Creation episodes from nearby Meso-america, emphasizing more themes of beginnings than of endings. One such scene comes from page 30 of the Borgia Codex, the Aztec-period almanac, a scene that may well represent the birth of time itself. The square-shaped frame around the edge of the page represents the body of a deity, perhaps an earth goddess, whose skull-like head can be seen at the upper center, with its arms and legs at the four corners. The scene within the frame is intended to be a womblike space, with the birth canal—an opening in the square—visible as a gap at the page's base. The interior circle is made up of radiating dark bands with eyes, perhaps representing the starry night sky, here depicted conceptually within the earth, and surrounds an image of wind gods and wind serpents inside the circle. Outside this circle we see a ring of twenty icon-like signs, many of them animal and bird heads. These correspond to the twenty day signs of the divinatory calendar, what the Aztecs called the *tonalpohualli*. Its first day is "Crocodile" (*Cipactli*), shown as the head of a fanged reptile within a small circle to the lower right of the scene. Three other small circles surround other day signs ("Death" [*Mictli*], "Monkey" [*Ozomatli*], and "Vulture" [*Cozcacuauhtli*]); all four are shown being pierced by a quartet of curious small figures who wield bone awls, each with a different plant or tree on its back. This shows us how four evenly spaced members of the list of twenty day signs correspond to one of the four cardinal points. My colleague Elizabeth Boone, an expert in the study of Aztec art and pictorial manuscripts, interprets these figures as earth spirit beings who, through ritual piercing, create and give life to the basic elements of time, already shown here anchored within its directional and cosmological framework. Time, like the other basic dimensions and properties of the world, was made inside the earth, which is also likened here to a nocturnal, primordial setting.[*]

The Aztec Calendar Stone

The most profound of all Mesoamerican representations of time and space is the famous Calendar Stone, or "Sun Stone," a large Aztec relief that now stands as a national symbol of Mexico and its pre-Columbian heritage. The Calendar Stone is perhaps most visible nowadays as a motif on tourist kitsch and billboards for Mexican beer,

---

[*] For the interpretation of this page of the Borgia Codex, see Boone, 2007, pp. 181–83. It isn't precisely accurate to say that the Borgia Codex and related screen-fold manuscripts of the so-called Borgia Group are "Aztec." Their content is closely related to the mythology and religion of central Mexico, but they were likely collected by the Spanish in the region of what is now Veracruz or Puebla, to the east of Tenochtitlan.

The Aztec Calendar Stone. (Drawing by Emily Umburger)

or as an oft-repeated symbol of "Maya-Aztec" intellectual achievement and mysticism. It happens to be profoundly misunderstood and misrepresented in many recent presentations, especially in writings and on TV shows concerning the Mesoamerican calendar and the approaching "end" of 2012. (The Aztecs, as we've seen, had absolutely nothing to say about the Maya date that falls in 2012, and probably weren't even aware of it.)

What's the real story and meaning behind this quintessential symbol of Mesoamerican time? The immense stone on which its image was carved was discovered in the year 1790, during digging of sewer lines along the great plaza, or *zócalo*, of central Mexico City. For months, workers had labored on the renovations in the area, and on August 11 of that year—a spectacularly fitting date, as we will see—they came upon a massive carved stone, today known as the statue of Coatlicue (Jade Is Her Skirt). The digging and renovation work continued in the *zócalo* after August, and on December 17 the great Calendar Stone was unearthed in a spot not far away. A few months later it was moved to the cathedral, where it remained for many years, visible to tourists and passersby, and today the stone is the centerpiece of the Sala Mexica in the National Museum of Anthropology in Mexico City.*

The surprising discovery of the demonic-looking Coatlicue figure and the Calendar Stone piqued the interest of the city's residents. By the 1790s a new era of enlightened thinking had begun to emerge in New Spain, and intellectuals and scholars in Mexico City were just beginning to be aware of pre-Hispanic history and culture. In earlier colonial times, Aztec sculptures discovered in what was once the great capital of Tenochtitlan were intentionally buried or destroyed outright; now, however, these ancient remains began to be seen more as historical curiosities, and as windows into Mexico's remote and cloudy past. The same central area of downtown Mexico City still yields Aztec treasures from time to time. In 1978, workers digging near a sewer drain (again) came upon a sculpture, this time of the goddess Coyolxauqui, which led to the discovery of the Templo Mayor, or Great Temple. Just recently, my friend Leonardo López Luján and his colleagues began work on the excavation of a massive offering deposit in front of the Templo Mayor; at this writing we still do not know what's inside, but no doubt there will be wonderful things.

The Calendar Stone was not a "working" calendar in the sense of being designed to be consulted as a tally of days, months, or years. Rather, the immense disc served a more simple purpose: it was a huge emblematic hieroglyph, representing the fifth Sun, *Nahui Ollin*, 4

---

* The original publication of the Calendar Stone by León y Gama was one of the first scholarly studies of a pre-Columbian monument. More recent presentations of the history of its discovery are by Matos and Solis, 2004, pp. 15–25; and López Luján, 2008; and Villela and Miller, 2010.

Earthquake. The design is wonderfully complex, but not quite as mystical or otherworldly as one might believe. In fact, it's quite "worldly" in a literal sense, for at the very center of it we see the menacing face of a man with a grimacing mouth, out of which emerges a tongue in the form of a flint-knife blade. This is the animate spirit of the surface of the earth, known as Tlaltecuhtli, or the Earth Lord. In Aztec iconography the Earth Lord often is recognizable by his toothy grin and the knife-tongue, surely a visual allusion to his ability to consume human flesh and blood through heart sacrifice. On the Calendar Stone, his face occupies the central circle of a large X-like image that is the hieroglyph for the day *Ollin*, "Earthquake." The four small circles attached to it give us a number four, providing the rest of the name for the current era of the universe. Tlaltecuhtli's face is thus an artistic elaboration on the day sign, which was itself a symbol of the earth; the corresponding day name in other Mesoamerican languages even means simply "earth."*

Four square panels around the visage of Tlaltecuhtli, still inside the *Ollin* symbol, provide other familiar time designations: 4 Jaguar, 4 Water, 4 Rain, and 4 Wind. These are, of course, the names of the four previous eras of the universe as described in the Legend of the Suns. The brilliant artist who designed the Calendar Stone visually conveys those earlier creations and destructions as parts of the current time, as necessary but incomplete components of the body of the world. Encircling the central *Nahui Ollin* glyph is a ring of twenty small squares, each containing an individual hieroglyphic sign. These are the glyphs for the twenty days comprising the 260-day divinatory calendar. The artist has

---

* Scholars of Aztec art and history have in fact debated the identity of the central face on the Calendar Stone. While I've clearly opted for its interpretation as the Earth Lord, others have long seen it as the face of Tonatiuh, the sun god. The debate is still unresolved, and it hinges on a simple visual question: does the face go with the outer sun disc, with its radiating points, or does it associate more directly with the *Ollin*, or "Earthquake," glyph nearer the center of the composition? I'll always defer to specialists in Aztec matters on this, but I am more comfortable with the "earth" (Tlaltecuhtli) option.

included them here as a circular frame for *Nahui Ollin,* conveying that "4 Earthquake" is a label for time as well as space, naming the sun that defines this current era of existence. The inner circle of days is, I think, an elaborate cartouche that visually conveys the brilliance and glory of the radiant sun. Repeating bands of turquoise and jade beads decorate its brilliant frame, as do the triangular points of light that radiate out from the center. The band of twenty days is the semantic "core" of a solar cartouche, showing that 4 Earthquake is a label for the Fifth Sun, and for time in a more abstract sense, as experienced in the current world.

The stone was carved during the reign of Motecuhzoma II (1502–1520) sometime in the last few years of the Aztec Empire. In its original setting in the central precinct of Tenochtitlan, this great circular monument was almost certainly set into the floor of a large outdoor platform, facing upward toward the blazing sun. I suspect that the remains of this platform may still exist under the place where the stone was first unearthed back in 1790. These sorts of raised surfaces were places of performance, often stage sets for rituals involving mock gladiatorial combat between captured warriors and elite guards of the emperor. Imagine, then, this elaborate carved stone as a horizontal surface, walked and bled upon as a living image of the sacred earth in the guise of its temporal identity 4 Earthquake, *Nahui Ollin.* As a public site of offering and sacrifice, the stone served as a symbolic and temporal place charged with cosmological meaning, providing the appropriate setting for violent reenactments of the sacrifice that created the world and everything in it. The circular design was a microcosm as well as a representation of the earth's "navel" opening, where the Tlaltecuhtli could receive his sustenance in the form of human blood.

Today, the Calendar Stone stands upright in one of the great museums of the world, where the face of Tlaltecuhtli looks out over the crowds of visiting schoolchildren and tourists, still surrounded by the layered emblems of time and space, but no longer able to convey its basic meanings as a horizontal symbol of earth and sun. Its original significance has been entirely stripped away. Today its function is to serve more as a symbol of national and cultural identity for Mexico.

So we see how the great Calendar Stone is a figurative portrait of

Aztec time, the current era of *Nahui Ollin*, coupled with an image of the face of the earth, of the world humans inhabit. This large day 4 Earthquake hieroglyph in the center of the design carried an obvious reminder of the mode of humanity's future destruction, through the movement of the earth and the wrath of Tlaltecuhtli. Unfortunately little is clear about this Aztec conception of their future, if it was ever firmly articulated through myth or prognostication. Personally, I believe that the era named 4 Earthquake, while "current" before 1521 and evoked by chroniclers of the sixteenth century, may not live on in the world of today. The Aztecs and their myths are largely gone, and the people transformed over the centuries into citizens of the Mexican nation. Perhaps the true "seismic" event of world destruction and transformation that ended the Fifth Sun, the age of *Nahui Ollin*, was the conquest of the capital of Tenochtitlan and its vast empire in 1521. The words of Aztec poets, sung and written just two short years after the victory of Cortés, offer a profound and emotional picture of a world swept away by an unstoppable force:

> *Nothing but flowers and songs of sorrow*
> *are left in Mexico and Tlatelolco,*
> *where once we saw warriors and wise men.*
> *We know it is true*
> *that we might perish,*
> *for we are mortal men.*
> *You, the Giver of Life,*
> *you have ordained it.*

> *We wander here and there*
> *in our desolate poverty.*
> *We are mortal men.*
> *We have seen bloodshed and pain*
> *where once we saw beauty and valor.*

> *We are crushed to the ground;*
> *we lie in ruins.*

There is nothing but grief and suffering
in Mexico and Tlatelolco,
where once we saw beauty and valor.

Have you grown weary of your servants?
Are you angry with your servants,
O Giver of Life?[2]

Maya Mythology

The image of the sun and cosmos represented on the Aztec Calendar Stone is often called "Maya" in books and TV shows, much to my chagrin. In fact the Maya had nothing to do with it. Such a loose label reflects a dismissive and woefully ignorant collapsing of two very distinct Mesoamerican cultures. To call the monument Maya would be similar to saying that the Roman Colosseum is Greek, or that Notre Dame Cathedral is Spanish. The Maya never made anything quite like the tableau of the Calendar Stone summarizing their view of the world's cycles of creation and destruction, and I'm sure its design would have made any Maya day priest scratch his head in confusion.

In fact, there's no similar corresponding Maya representation of the cosmos. Instead we have a wide variety of different sources to draw upon, ranging from ancient and historical texts to scenes of iconography, all produced over the course of some two thousand years. Sometimes they're even contradictory with regard to one another. It makes one realize that there is no single Maya Creation story or myth that we can simply pull off the shelf and ponder. Looking at the ancient sources especially, we see that the Creation was very local in its flavor, with different kingdoms and cities claiming different supernatural origins, each a center of the cosmos in its own way. This, too, was a very widespread Mesoamerican idea, perhaps not so visible among the Aztecs because Tenochtitlan, as the center of an empire, was hardly just another Mexican kingdom; as the imperial capital, its cosmos was the world's cosmos.

One centerpiece of Maya mythology as we understand it today is a remarkable account of world creation and heroic exploits known as the Popol Vuh, or the "Community Book." Preserved in the form of a colonial manuscript written in an archaic version of the K'iché' Mayan language, it relates how the cosmos began, how the moral universe arose through the defeat of the Lords of Death, and how humanity was thus created. While truly cosmic in scope, the Popol Vuh, as its name implies, was also very much a local product, rooted in the history and ethnic politics involving the K'iché' kingdom in what is now highland Guatemala. Much of the work is devoted to the exploits of the K'iché' kings who ruled at Cumarcah, their capital, who, in a very explicit way, are likened to the cosmic heroes of the primordial past. So, while it might be easy to consider the Popol Vuh as something akin to a Maya Bible, it strongly reflects the real political and social landscape of the K'iché' Maya on the verge of the Spanish Conquest. It's a work that strives to connect primordial myth with the royal dynasties of the time, following a pattern that Maya literature followed for generations, long before European contact.

The Popol Vuh's origins are obscure, and much of its substance is no doubt quite ancient. The document's very preservation is something of a miracle, in fact. It comes down to us only in the form of a secondhand copy diligently made by one Francisco Ximénez in around 1704, when he served as a Dominican monk in the highland Guatemalan town of Chichicastenango. Ximénez was a remarkable scholar in his own day, having learned a number of Mayan languages, including fluent K'iché', the language of the Popol Vuh document. Generations earlier, Spanish priests would most likely have destroyed such expressions of indigenous belief, but Father Ximénez, living in a somewhat different and more enlightened time, clearly saw its cultural and historical significance and worked hard to preserve it. The tattered manuscript he found in Chichicastenango was already very old, probably dating to the first decades after the Spanish Conquest of the K'iché' in the sixteenth century. That first copy, in turn, was very likely an alphabetic transcription of a long-lost pre-Columbian document, probably a screen-fold pictorial work similar in general appearance to the codices from highland Mexico, such as the Codex Borgia.

Ximénez made his copy of the original K'iché' text and then crafted a direct Spanish translation, and he may well have prepared secondary copies to distribute to other archives. One of these manuscripts came to light in the mid-nineteenth century, in the fledgling years of Maya academic research, and saw publication in French in 1861. The manuscript upon which that was based, the oldest known copy of the Popol Vuh, probably in Ximénez's own hand, was eventually obtained by the Newberry Library, in Chicago, where it now resides.

The Popol Vuh is perhaps the foremost example of Native American literature, giving voice to ancient, compelling notions of world creation, or what its authors called "the completion and germination of all the sky and the earth."[3] The opening lines of its section devoted to the primordial world give a sense of the document's epic scope:

This is the account of when all is still silent and placid. All is silent and calm. Hushed and empty is the womb of the sky.

These, then, are the first words, the first speech. There is not yet one person, one animal, bird, fish, crab, tree, rock, hollow, canyon, meadow or forest. All alone the sky exists. The face of the earth has not yet appeared. Alone lies the expanse of the sea, along with the womb of all the sky. There is not yet anything gathered together. All is at rest. Nothing stirs. All is languid, at rest in the sky. There is not yet anything standing erect. Only the expanse of the water, only the tranquil sea lies alone. There is not anything that might exist. All lies placid and silent in the darkness, in the night.[4]

The Creation took place when the two gods "the Framer and the Shaper" gathered in the primordial waters before the most prominent Creator deity of all, named Heart of Sky. Together they pondered the beginning of the world, and through the power of words gave substance to the cosmos: "In order to create the earth, they said 'earth,' and immediately it was created." The gods created all of the animals of the world, but they were crestfallen to find that the animals could speak properly; the beasts "only squawked and chattered and roared," unable to venerate the Creators. Their creations, they soon realized, meant little without the presence of humans to venerate the gods for their powers and abilities. The gods said to one another, "There can be no worship, no reverence given by what we have framed and what we have shaped, until humanity has been created, until people have been made."[5] So, in their next attempt, they fashioned people out of mud. This, too, failed, as the mud people soon fell apart and crumbled in water.

Frustrated again, the Creator couple known as Framer and Shaper turned to divination in order to perfect the creation of humans. They performed a ceremony using beans or maize grains, counting the days of the sacred 260-day calendar, just as it is still performed by modern-day diviners in the Maya highlands. This "revelation of days and of shaping" shows that the introduction of calendrical ritual was necessary to ensure a proper form of humanity.[6] It was still not enough. The next effort by the gods involved the carving of humans out of wood, but these humans lacked hearts and minds. "They walked without purpose," and "were merely an experiment, an attempt at people," devoid of movement and life-giving blood. Heart of Sky conjured a great flood, and most of the people of wood, like the mud people before them, were destroyed. Their few descendants resided in the world as spider monkeys, vaguely appearing like people.

The Popol Vuh's story then pauses to relate the story of two pairs of heroic figures who play essential roles in defeating certain negative forces in the primordial world. The first pair is named One Hunahpu and Seven Hunahpu—both calendar names in the 260-day system—twins born to the Creator couple named at the beginning of the epic. The young men grow to enjoy a life of leisure, becoming master ballplayers and passing much of their time playing dice. The elder, One Hunahpu, has two sons of his own, named One Batz' and One Chouen, who become artists, musicians, and craftsmen. Together they all often play an outdoor ball game, a version of the sport played throughout ancient Mesoamerica, making use of a large rubber ball that bounces between individuals or teamed contestants. Their games are very noisy affairs, disturbing the gods who preside in the Underworld, in what the Maya called Xibalba', "Place of Fear." The Lords of Death are appalled by the ruckus and by One Hunahpu and Seven Hunahpu's lack of respect for them. They challenge the twins to a contest in Xibalba', hoping to defeat them and also to acquire their fine ballplaying gear, including arm protectors, face masks, and headdresses, "the finery of One Hunahpu and Seven Hunaphu." The twins are summoned to the Underworld, first reassuring their mother that they will be safe.

One Hunahpu and Seven Hunahpu descend into an eerie world of canyons and horrifying streams and rivers, where scorpions live and blood flows in place of water. At last they arrive at Xibalba', at the court of the Underworld lords named One Death and Seven Death, who are intent on sacrificing the young ballplayers.* The next day, the deed is carried out, and One Hunahpu and Seven Hunahpu are killed. The lords of Xibalba' decree that the head of One Hunahpu be put in a barren tree by the roadside; when the head is placed there, the tree suddenly bears fruit, with the head itself transforming into a calabash.

One day soon thereafter, a young maiden named Lady Blood, daughter of one of the Xibalba' rulers, is walking along the road and she goes up to the great tree, having heard of its magical fruits. As she approaches, the skull of One Hunahpu (not completely a calabash, one supposes) speaks out, calling her to stretch out her right hand. One Hunahpu drops saliva into Lady Blood's palm, impregnating her with twins. He tells the surprised girl to leave Xibalba' quickly and to go up

---

* The underworld, Xibalba', as described in the Popol Vuh, consisted of a number of harrowing locales named the House of Darkness, the Shivering House, the Jaguar House, the Bat House, and the Blade House. I suspect that these five dangerous places were arranged in Xibalba' in the familiar four-around-one format of the quincunx, the five-part diagram of the cosmos found throughout ancient Mesoamerica.

to the surface of the earth, where she will find One Hunahpu's mother and receive proper care. First, however, she has to deceive her own father, who is livid that she will have a child out of wedlock. She makes her way to the surface of the earth, where she eventually gives birth to a new set of heroic twins, named Hunahpu and Xbalanque, and it is they who become major actors in setting the stage for the dawning of a new world where fully formed humans can exist and thrive.

Before this can happen, however, the world, still dark and incomplete, has to put up with a great, proud, and thoroughly annoying celestial bird named Seven Macaw. (I wonder if the loud and nettlesome personality of the macaws I know is reflected in the personality of this deity.) In the primordial darkness still overseen by Heart of Sky, during the era of the wood people, the bejeweled Seven Macaw boasts that he is the true light of heaven, the essence of the sun and the moon, who can "light the walkways and pathways of the people."[7] Hunahpu and Xbalanque, accomplished hunters by this time, become fed up with Seven Macaw's egotistical behavior and are determined to shoot him down from the sky using their blowguns. As the great bird sits in a large fruit tree, the twins shoot him down and defeat him, but not before Seven Macaw wrenches off Hunahpu's arm in retaliation. The bird's gold and greenstone finery is stripped away, and Seven Macaw is properly humbled, paving the way for the true source of light in the later Creation.

The importance of the Popol Vuh in Maya scholarship is difficult to overestimate, for obvious reasons. We know parts of the story are quite old and found in many regions of the Maya area, despite the local spin of the larger narrative as it exists today. For example, antecedents of the story, or elements of it, are clearly reflected in ancient Maya art as remote in time as the late Preclassic, around 100 BC. One sculpted monument of that era, from the early Maya center of Izapa, on the Pacific Coast, depicts a large, fantastical bird perched atop a pole near a tree, above two men, one of whom is missing an arm. There can be little doubt that this avian creature, whom we call the Principal Bird Deity, is equivalent to Seven Macaw of the Popol Vuh, and that the small men are Hunahpu and his brother. Similar images of the twins shooting the great bird with blowguns are known from later Classic-era Maya ceramics dating several centuries after Izapa (see page 300). Incredibly, these images reveal that this episode in the Popol Vuh story can be traced back some sixteen centuries before Ximénez copied the old, tattered manuscript in Chichicastenango.

Curiously, apart from a handful of representations of the twin hunters and the great bird, very little else of the Popol Vuh has strong parallels in ancient Maya art or mythology. Nothing on stone monuments or ceramics depicts the episodes of early world creation by Framer and Shaper, nor do we see any clear depiction of the twins' father and uncle, One Hunahpu and Seven Hunahpu, in their ballgame and epic battle with the Underworld lords. In fact, looking at the entire epic story presented in the Popol Vuh, one gets the sense that, at the time of its discovery by Ximénez, it was a living document, closely guarded by the elders of the community but also firmly reflecting the time of its creation in the K'iché' highlands of the early sixteenth century. It certainly incorporates old elements and old stories, but it weaves them as a backdrop to what amounts to its real purpose: a political narrative affirming the position and status of the K'iché' nobility on the cusp of their destruction at the time of the Conquest.

Despite all of this, there has long been a strong tendency in Maya scholarship to see the Popol Vuh less as a document from a specific time and place and more as a vivid survival of an ancient, Classic-era Maya vision of religion and politics. Some of my colleagues may not like my saying this, but I'm of the strong opinion that the Popol Vuh has been overused as *the* document on Maya religious thought, almost as a template or lens through which we can interpret much of the ancient culture's art and cosmology. But as the deciphering of ancient writing now reveals, ancient Maya cosmology possesses its own internal body of texts, closer in time and place for interpreting the details of iconography and ancient mythology. Often these more ancient stories, depicted on painted ceramics or in ancient myths, are quite different from the narratives of the Popol Vuh, and even show vignettes of mythology cycles now altogether lost.

The Stela and the Hearth

If we were to choose one principal "Creation" date from Classic Maya mythology, it would be the Long Count 13.0.0.0.0 4 Ahaw 8 Kumk'u, corresponding to August 11, 3114 BC. Now, it may not be completely accurate to say that this was a true Creation date, because ancient religious texts tell of events and episodes that took place long before this date—sometimes millions of years before, in fact. But it is fair to say that the ancient Maya must have considered 4 Ahaw 8 Kumk'u to be the beginning date of our current era, when, according to some sources, the gods of the cosmos "were set in order." With that, all else was possible.

This 4 Ahaw 8 Kumk'u date is explained in considerable detail in the inscription of a tall monument erected at the ruins of Quiriguá, Guatemala, known to archaeologists as Stela C. It was erected by a local king named K'ahk' Tiliw Chan Yopaat to commemorate a station in the Long Count, a period ending on December 23, AD 775, very close to winter solstice. There are two narrative texts inscribed in the narrow sides of sacred stone, one mythical, relating the events of the Creation in 3114 BC, and the other historical, mentioning an earlier king of Quiriguá who performed his own period-ending rite on August 25, AD 455, a little over 320 years earlier.

The left side begins with an Initial Series date, a Long Count, written as 13.0.0.0.0 4 Ahaw 8 Kumk'u. A record of the events of that momentous day comes in the glyph blocks that follow (see page 218). In the first we find a small *X*-like element, which we know from Mayan syntax represents the core or "root" of a verb. There is some evidence, not completely secure as yet, that this *X* sign reads *jel,* a word in Mayan languages that might sound familiar from earlier discussions in chapter 3, meaning "to switch" or "change-over."

What changed? That is what we see in the next glyph, the subject of the verb. Here we have a sequence of two syllables, *k'o* and *ba.* In a parallel case, the same glyph looks to be spelled *k'o-jo-ba,* suggesting that the Quiriguá glyph is a shorted form of the word *k'oj ba,* perhaps pronounced and spelled sometimes as *k'o-ba.* The best reading for

Stela C from Quiriguá, Guatemala. (Photograph by the author)

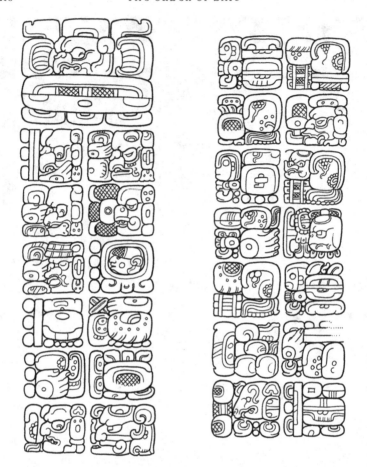

The account of Creation in the inscription from Quiriguá's Stela C.
(Drawing by the author)

this word, I believe, is based on the root *k'oj,* meaning "image, mask." In Yukatek Mayan, the equivalent is *k'ohbail,* "image, form, portrait," and in Ch'ol Mayan the equivalent is *k'ojbäjil,* with a somewhat related meaning of "head, face."* Both of these are in turn related to the

---

* The Ch'ol term comes from an early vocabulary by Hopkins, Guzman, and Josserand, 2008.

widespread Mayan word *k'oj,* meaning "mask." The idea of these obviously related concepts centers on the key notion that one's head or face is what conveys one's image and identity. The larger glyphic phrase therefore may well mean something like this:

jehlaj k'oj baah,
the face-image changed

Rather enigmatic, to be sure, and we'll come back to it. Next comes a glyph with the number three, but it's not a date. It's another sentence that reads:

ux k'ahlaj tuun
thrice are bound(?) the stones

This refers to an important ceremony of stone dedication, common in history during a period ending, when the twenty stones (*tuuns*) of a K'atun or its subdivision were gathered and bound together as a mark of its completion. Here, three stones were similarly set together in primordial time.

According to the glyphs that follow, these three stones were thought of as three different monuments, probably upright stelae, dedicated by a pair of rather enigmatic deities we know of as "the paddlers." (In one of their few portraits, they are shown rowing a canoe bearing the maize god and several animals, clearly a scene from a myth now lost to us.) In numerous other ritual texts the same paddler gods are said to watch over the calendar ceremonies of the kings, sanctifying the ritual through a metaphorical "bathing" of time and its celebration. Stela C shows that their important roles as godly patrons of period endings in historical time was a continuation of their much older mythological tasks as participants of Creation. The three monumental stones carry names that still elude decipherment, but each appears to be classified in a different way: One is a "jaguar stone," planted into the ground by the paddler gods in a place known perhaps as the "Five Sky House" or "Five Sky Houses." The second is a "snake stone," planted by a god whose name, frustratingly, appears nowhere else in Maya mythology,

The third stone is called a "water stone," and this was "bound" with the others, perhaps, by the great celestial god we call Itzamnaah, who ruled the heavens from a cosmic throne in the sky. The idea of binding means that this was the completion of the set, in some sense the fastening together or gathering of the individual stones. The text concludes by telling us that the overall occasion of Creation and the setting of the three stones were overseen by the "Six Sky Lords." We do not know who these deities were, but the number six suggests a solar arrangement of some sort, perhaps referring to the four solstice points along with zenith and nadir. Alternatively, it might relate to the fact that the dedication date of Stela C fell on the day 6 Ahaw (9.17.5.0.0 6 Ahaw 13 K'ayab).

Two glyphs provide important information about where this "image change" happened:

ti' chan, Yax Yoket Nal(?)
(at) the edge of the sky, (at the) New Hearth Place(?).

This final place name has received a good deal of attention from Mayanists in recent years. As Linda Schele points out, the main component of its glyph represents three stones arranged in triangular fashion, surely a visual cue to the three rocks that make up a hearth in any traditional Mesoamerican house (what is sometimes called a *tenamaste*). In modern Mayan languages close to what we know was spoken by ancient scribes, the word for the three hearthstones is *yoket,* which can be a very provisional reading in this setting.

Stela C is not the only text to refer to these ideas of "image change" and hearths. In fact, the notion is found throughout the Maya region at many different sites of the Classic period, usually distilled to its essentials, as in this phrase from a text at Palenque:

jelaj k'ojbah ti' chan Yax Yoket Nal(?)
The face/image changed at the sky's edge, at the First Hearth Place.

I think that we'll be debating the precise meaning of this opaque phrase for a long time. At the very least it refers to some type of replacement or switch from a previous state (*jel*) involving "images" or "masks," at or near the "First Hearth."

In their influential book *Maya Cosmos*—the first work to offer a coherent interpretation of this Creation event—Linda Schele and her colleagues advocate for an astronomical interpretation of the phrase, which she translates a bit differently than I have here.* Schele links the "First Hearth" reference to the modern K'iché' Maya belief that a triangular arrangement of three stars in the constellation Orion (Rigel, Alnitak, and Saiph) represents a hearth, its "smoke" being the Orion Nebula, visible in the center of the triangle. In making this association, she proposes that the date of Creation was the centering of the cosmos by the gods who dedicated the three stellar hearthstones in Orion on 13.0.0.0.0 4 Ahaw 8 Kumk'u. Schele and her colleagues go much further, making a number of related astronomical readings of mythical events cited in the art and inscriptions at Palenque and other sites.

I remain somewhat skeptical of many of their astronomical interpretations, to be honest (I explore these ideas further in chapter 10). The reading of the three stars of Orion as a hearth arrangement seems reasonable, perhaps, but it simply grafts modern K'iché' ideas onto the mythology of the Classic Maya. Significantly, other Maya of today do not see a hearth in that location, which raises a number of questions and doubts in my mind. We'll need far more than a simple one-to-one parallel to make a firm interpretation of the still obscure contents of classic Maya Creation mythology.

Another possibility worth considering is that the three stones of the Stela C text refer to the establishment of a geometric ideal that's reflected in several different kinds of triadic arrangements and dimensions in Maya cosmology. As we've seen, three-part divisions of space are common in the Mesoamerican worldview, most obviously in the fundamental division of the world into sky, earth, and Underworld. I'm

---

* For a detailed analysis of the Stela C text, see Schele, Freidel, and Parker, 1993, pp. 59–122, and especially Looper, 2003, pp. 158–85. While I differ from Looper in some details of his glyph analysis, his presentation is nuanced and particularly insightful.

reminded also of the basic three-part movement perceived in the coursing of celestial bodies across the sky, emerging from the east, running
above the earth at noon, and setting in the west, what K'iché' Maya of
today call "the three sides, the three corners."[8] In other words, we need
not link the hearth of Maya Creation mythology to a specific constellation of stars. To me, the sense is more general and overarching, clarifying that on the day of Creation, the gods established a three-part
dimension to the universe, complementing the other sacred numbers,
such as four, five, thirteen, and twenty, reflected in the structures of
the cosmos.

Returning to the textual description of the episode, just what is
meant by the "changing of the image" or "changing of the mask"? This
is difficult to know. Perhaps the word *k'oj* refers to masks, images, or
faces that should be equated in some manner with the three sacred
stones dedicated on that day by the gods. I suggest this as a possibility
because we've long known that three stone heads or masks along a celestial band comprise an important cosmological symbol for the Classic
Maya, most often manifested as small portrait heads attached to
"sky-belts" worn by Maya kings as part of the ceremonial costume for
period-ending rituals. The "change of masks" might, then, refer to the
idea of the cosmos getting a new identity of some type—a makeover of
sorts—which in turn became symbolically reflected in the ritual dress
of Maya kings, and especially in their cosmic belts.

The nature of these seminal events related to the Creation base date of
3114 BC—the stones, the "images," the hearth, and so on—will be studied
over and over for the foreseeable future. Mythology is one of the more
exciting aspects of ongoing work in Maya decipherment and art history,
where confident pronouncements of a final answer or a complete understanding are always going to be met with some skepticism and doubt.

The Gods Align

Along with Stela C, the greatest source for understanding this Creation
date is the beautifully painted vessel known as the Vase of the Seven

Gods. It was originally painted by an artist who was a member of the royal court of Naranjo, a major kingdom in what is today northern Guatemala. Like many pottery vessels from Classic Maya culture, the tall, cylinder-shaped vase was used as a ceremonial container for a fermented chocolate drink, a function explicitly described in the horizontal band of glyphs above the scene (one of the glyphs reads *ka-ka-wa* to spell the familiar Mayan word *kakaw*, "chocoloate").

Dominating the painted scene on the right is an enthroned god who wears a large brimmed hat decorated with bird feathers. He is an old jaguar deity known to Mayanists as "God L," surely related in some way to the jaguar god who serves as the animate aspect of the night sun. He sits on an elaborate jaguar seat, accompanied by large round bundles probably containing jade beads and other riches. He gestures to six other seated gods, divided into two tiers. A hieroglyphic text runs down the middle of the scene, describing the setting. The first glyphs record the Calendar Round date 4 Ahaw 8 Kumk'u—a state-

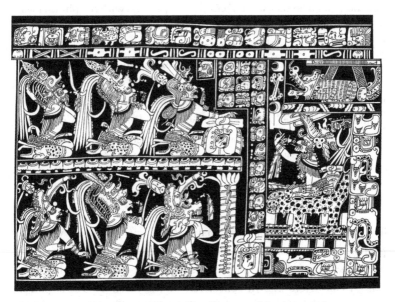

Maya Creation as Depicted on the Vase of the Seven Gods.
(Drawing by Diane Griffiths Peck)

ment sufficient enough to establish that we are at the time of Creation, on 13.0.0.0.0 4 Ahaw 8 Kumk'u. The third glyph in the sequence is the verb, reading *tz'ahkaj*, "(they) are ordered, aligned." The subjects of the verb make up nearly the entire rest of the text, named in sequential pairs of hieroglyphs, each of which includes the word *k'uh*, "god." There are six such pairings in all, and their listlike presentation suggests that six collections, or sets, of deities are all ordered and set in place on the day of Creation. Two of the categories of supernatural beings are simply described as "the heavenly gods" and "the earthly gods." Another unclear term is the name *Bolon Yokte' K'uh*, "the Nine Poles Gods." We will later find this used in connection with the single Maya record of the 2012 anniversary of Creation.

Another painted chocolate vessel is nearly identical to this Vase of the Seven Gods, showing a few more gods thrown into the mix—we know it as the "Vase of the Eleven Gods." Apart from a more populated scene, its imagery and texts are nearly identical to the other. One key difference is that the vertical text mentions not only the "when" and "what" of Creation, but also the "where," closing the passage with the phrase *uhtiiy K'inchil*, "it happened at K'inchil." This is the mythical location meaning "Great Sun Place," perhaps referring to the realm of the sun god. Taking the imagery and texts as a whole, these extraordinary vessels seem to reference the initial ordering of broad categories of divine beings in what looks to be a dark, cavelike setting, not the brilliant throne room of the animate sun, K'inich Ajaw. Perhaps we are to believe that K'inchil, the sun's locale, was still in primordial darkness when Creation occurred.

The Palenque Mythology

The best lengthy account we have of an ancient Maya vision of primordial time comes from Palenque, written on stone tablets that decorated three of its major temples. These story lines involve three central characters known collectively as the Palenque Triad, patron deities and remote ancestral gods of the local ruling dynasty. Their full actual names remain obscure, but Mayanists have long referred

The Group of the Cross at Palenque. (Photograph by the author)

to them individually as GI, GII, and GIII (*G* here standing for *God*). The temples dedicated to them still stand, tucked beneath a beautiful green hill at the edge of the site's center, in an area of the ruins known as the Group of the Cross. They were all dedicated as a set in the year AD 692, early in the reign of the Palenque king named K'inich Kan Bahlam. According to the texts in the temple shrines, the Triad were siblings in a sense, all born within a few days of one another in the distant past, not long after the creation of the world in 3114 BC.

Chief among them was GI, a solar deity who seems also to have had close associations with the ocean and primordial seas. His shrine is the tallest of the set, known as the Temple of the Cross, and its symbolism is clearly tied to the celestial realm. GI himself is an aquatic being; most likely we think of him as a watery aspect of the sun god K'inich Ajaw. The connection between sky and water may be based on GI's specific role as a symbol of the predawn sun, the solar aspect that exists within the eastern ocean, and that ascends into the heavens to bring the light of day. I suspect that GI is specifically the god of the sunrise, and that his "house" is at the center of the sky, as a manifestation of the sun's zenith. Another important theme of the temple is royal ancestry, as shown by the historical narrative inscribed on its main tablet, including a long list of the births and inaugurations of numerous early Palenque kings.

GII's shrine, the Temple of the Foliated Cross, is slightly less tall than the Temple of the Cross, and its symbolism emphasizes a watery surface of the earth, replete with ripe maize plants. This god's name we actually know, since so far only his name hieroglyph is readable. He was Unen K'awiil, "Infant K'awiil," in reference to a youthful aspect of K'awiil, one of the most important gods or spirits of ancient Maya religion. K'awiil was a complex entity who symbolized many different things at once, all revolving around royal power, lightning, ancestry, and concepts of procreation. His specific infant aspect is a bit difficult to grasp, but we find that "baby gods" do have parallels elsewhere in Mesoamerica. Among the modern-day Ch'ortí Maya of Guatemala, for example, the infant form of the god of maize is vitally important as a symbol of the new corn of the cornfield. At Palenque, Unen K'awiil appears to serve as a key symbol of earth and agricultural fertility.

The god GIII, like GI, seems to be an aspect of the sun god, but is very different in his symbolism and associations. He is the nocturnal sun, sometimes labeled by modern researchers (not by the ancient Maya) as the "Jaguar God of the Underworld." His shrine in the Temple of the Sun is more correctly a temple of darkness, symbolically evoking the interior of a sacred cave. It stands in symbolic opposition to the daytime solar associations of the Temple of the Cross, and was dedicated to themes of warfare and sacrifice.

The three temples are arranged on the hillside much like a hearth, and I'm certain this was intentional. All three were built by K'inich Kan Bahlam in anticipation of the great k'atun ending of 9.13.0.0.0 8 Ahaw 8 Woh, on March 13, AD 692, and were meant to serve as a figurative hearth, re-creating the sacred space of Maya Creation. In fact, the Creation base date and mention of the *Yax Yoket Nal*(?), "First Hearth Place," is cited several times among the three temples. The symbolism is even more profound in light of the period ending itself: thirteen k'atuns, a very special occasion in Maya history. The Group of the Cross is at once a shrine dedicated to local dynastic gods and a reproduction of the triadic arrangement set at the time of Creation. Even today, more than thirteen centuries after they were built, these remarkable temples remain fairly well preserved, and for me they retain much of their special aura.

Similar trios of gods appear at other ancient kingdoms, such as Caracol, Tikal, and Yaxchilan, but their individual members vary from place to place and are never consistent. I nonetheless suspect that each set of deities symbolically evokes the same basic Creation myth, linking them to the hearthstones of Creation, and perhaps to the idea of a "change of masks" or "images." Palenque's role in the study of Maya myth will always be special simply because of its well-preserved art and inscriptions, but I suspect that similar three-part temple plans were common in other Maya centers, and are still to be found through excavation and further research. In fact, the design of some of the largest temples of the Preclassic era show a clear triadic design that's eerily similar to what we see in Palenque's Group of the Cross, perhaps derived also from the arrangement of a mythical hearth.

As in politics, mythology, for the Maya at least, is local. At the same time, though, it is remarkable how universal most elements of Maya myth cosmology are, at least in the Classic period. Important gods were venerated across the landscape and over centuries of history and political and social upheaval. We see consistent notions of a cosmic architecture, and mythological events, places, and characters appear throughout the inscriptions of many widely spaced centers. There's a fair likelihood, then, that Maya culture of the Classic period held to a fundamentally shared cosmology, but one that could accommodate any number of localized stories, characters, and elements.

Conspicuous in its absence in any of the written records at Palenque or elsewhere is any mention of what comes next at the opposite of Creation, at the end of everything. Given all of the current interest in how the ancient Maya are said to have "predicted" an end to the world, or how they knew that their calendar would "end" in 2012, this might come as something of a surprise. The truth is, *no Maya text—ancient, colonial, or modern—ever predicted the end of time or the end of the world.* Perhaps now you can begin to understand my frustration at all the silly nonsense we hear as the year 2012 approaches.

If 2012 isn't an "end," then what is it? We ought to look a bit more closely at the Long Count calendar to see how 2012 isn't just another period ending or turn of a bak'tun. As we'll see in the next chapter, it's a very special date indeed, when, thanks to some tricky Maya math, we find ourselves in what the Yukatek Maya called a "fold" in time. I can't be certain what it all means, but at least now, for the first time, we can begin to understand what the Maya truly thought about the year's importance in the grander scheme of cosmic time, well beyond what some falsely claim will be "the end of times."

8

THE DEEPEST TIME

*This is the history of the World in those times, because it has been
written down, because the time has not yet ended for making these
books, these many explanations, so that Maya men may be asked
if they know how they were born here in this country, when the land
was founded.*

—*The Book of Chilam Balam of Chumayel*[1]

Our own cosmology, as accepted by most scientists, posits that
the universe came into being nearly fourteen billion years
ago. Space and time did not exist before this initial moment,
when all of the universe's matter and energy, condensed into an infi-
nitely dense, hot, and small "singularity," suddenly began its massive
expansion and cooling. Subatomic particles soon formed, and from
these the basic light element of hydrogen. Through the process of fusion in the dense young universe, and in young stars, many heavier
elements came into being, including the all-important carbon atoms
so basic to organic chemistry and the advent of life. Earth's formation
some 4.5 billion years ago, in the last third of the universe's age, has its
own complex story. A billion or so years into its history, amino acids
were somehow able to form into proteins, and through organic chemi-
cal processes still debated, molecules were able to self-replicate and
diverge in form. Life began to evolve.

That's it, in an overly simplistic and vague nutshell. And it is indeed
vague because no one really knows just how it all happened. We have

reasonable theories for what happened at the beginning of time and space, and maybe how the earth and life on earth arose much later, but in reality, nobody can really explain any of it in much detail. So here we all are: human beings farming the soil or, in my case, typing at a computer; house cats mewing for attention; trees growing and being cut down; mosquitoes landing on the arms of sweaty humans looking for ruins in the vanishing jungle. Everything "out there"—all life on earth, all that we can perceive around us, and all the vast matters in the universe—has come into being in the last 13.7 billion years. We may not understand much about how it all happened, but at least we have a pretty good idea of this cosmological time frame for the events and developments I've described so superficially, involving truly incomprehensible spans of time.

The scale of our "deep time" cosmology pales in comparison to the temporal framework with which the ancient Maya conceived of their universe. Believe it or not, our cosmos of 14 billion years—let's round the number up to be generous—is not even 0.00000000000000005 percent of the Maya notion of how long the universe will last. As we will see, they conceived of the beginning of time as having occurred 28 octillion, 679 septillion years ago. Now we're talking incomprehensible. While it doesn't agree with our imperfectly reconstructed cosmology, this truly staggering amount of time, indicated in a handful of calendrical records from the Classic period, represents as far as I'm aware the longest single conception of time in human history. Maya time, it's fair to say, is humanity's "deepest."

Numbers upon Numbers

The standard Long Count calendar has five units, the largest of which, the bak'tun, roughly equates to 400 years. This provided a perfectly adequate mechanism for tracking historical time, but it was structurally limited when it came to recording and computing very large numbers of years. The largest possible number one can represent in the standard Long Count system, maximizing the coefficients on each of the five units, is 19.19.19.17.19, equal to 2,879,999 days, or over 7,885 years. In certain ritual or mythical texts, Maya scribes felt the need to compute greater quantities of time—sometimes very much greater—so the standard Long Count simply wouldn't do.

In order to stretch out the scale of time, the Maya therefore made use of various time periods *above* the bak'tun, the highest of the standard five units we've discussed so far. In fact, it so happens that the standard five-part Long Count is actually a truncated version of a far, far larger system. This I prefer to call the Grand Long Count, to distinguish it from its more customary shortened form, where the bak'tun is the largest unit. The standard system represents the last five positions in this vast cyclical arrangement we'll now explore in some detail.

The existence of large calendar periods above the bak'tun has been known to Mayanists for many years. In the late nineteenth century, Förstemann didn't come across any of them in his work with the Dresden Codex, but somewhat later, in the 1920s and '30s, other early specialists in Maya glyphs studies were intrigued to find a handful of examples of such higher elements scattered among a few obscure inscriptions. Such efforts, spearheaded by Sylvanus Morley in those early years, led to the identification of periods known as the piktun (20 bak'tuns), followed in turn by the kalabtun (20^2 bak'tuns), k'inchiltun (20^3 bak'tuns), and alawtun (20^4 bak'tuns). (Again, these names were more or less coined by

"PIKTUN"
(20 Bak'tuns)

"KALABTUN"
(20 x 20 Bak'tuns)

"ALAWTUN"
(20 x 20 x 20 Bak'tuns)

Three of the "high" periods of the Maya Grand Long Count.
(Drawings by the author)

Morley; they were never used by the ancient Maya.)* Just as with the tun, k'atun, and bak'tun, each of the higher periods in the base-twenty vigesimal system is exponentially higher than the last, so when we consider the alawtun, for example, we're dealing with a very, very big unit of time, each lasting roughly 64 million years.

Throughout most of the twentieth century the rare alawtun was the highest-known period in the Maya calendar. Other, still higher units were mathematically conceivable, but poorly preserved examples of them made their study and identification all but impossible. In the 1980s this began to change in an important and surprising way when my colleague Linda Schele and I began analyzing in detail a handful of unusual Long Count dates that went far beyond the time scale of even the alawtun. What we realized was that the alawtun period of sixty-four million years turns out to be just one of an enormous array of increasingly larger units, and therefore not at all close to conveying the full extent of Maya time computation. In fact, some of the highest periods of the system truly dwarf even the alawtun.

To see Maya time in its full glory, we need to turn to two very eroded limestone slabs that still stand upright in the jungles near the ruins of Coba, where I spent a good deal of time as a youngster, accompanying my parents on their expedition to map and record the thousands of buildings scattered in the rain forest. In the Maya village surrounding Lake Coba we all lived for many months crowded in a modest thatched hut. It was there that I participated in the Ch'a Chaak rain ceremony in the dry summer of 1975. Little did I know then that Coba would hold an essential piece of the puzzle for our understanding of the structure of Maya time and cosmology.

As an archaeological site, Coba is hard to beat (though I admit I'm biased). Its architectural remains are found in clusters throughout the forest, over an area of roughly fifty square kilometers. Long, straight causeways radiate out from the site's core area nestled among a series of large, beautiful lakes. One of these ancient roads runs for more than

---

* Morley, for example, was the first to discuss the higher periods, calling the piktun the "Great Cycle." See Morley, 1915, pp. 117–18.

sixty miles through the dense forest, linking Coba to another large ruin to the west, Yaxuna, just to the south of Chichen Itzá. It's the longest ancient Maya road known. Very little of Coba's ruins have been excavated, and in fact the site is so vast that it will probably never be adequately explored and investigated. Most of the many buildings and pyramids at the ruins date to the Classic period, as best as we can guess, and the site appears to have been abandoned by the Postclassic era, around AD 1000. Coba was revived as a major ritual center during the final years of pre-Columbian history, in the late fifteenth and early sixteenth century, when many of its old temples were refurbished and its many carved stelae were reset. This drastic renovation of the site suggests that the Maya of the fifteenth century or so were attempting to revive the glory of an ancient capital, reusing ancient monuments that were already centuries old.

In its heyday, during the seventh and eighth centuries, Coba was no doubt at the center of an important political and ceremonial network, and it may have been a significant pilgrimage center for much of the surrounding population throughout northeastern Yucatán, if not beyond. Unfortunately, we know little about Coba's rulers or the town's dynastic history, due to the terribly eroded condition of its many sculptures and inscriptions. But there can be little doubt it was one of the greatest of all Maya cities. Today one can still walk in any direction in the forests around Coba and immediately encounter endless mounds and raised roadways. If one ever wants to experience a sense of old-time exploration in the jungle, Coba is the place to go.*

Coba has many broken and poorly preserved stelae, two of which are of extreme importance for our understanding of the full structure of the Maya calendar. These are known as Stela 1 and Stela 5, both still standing erect in the Macanxoc area of the site. Stela 5 was erected on August 20, AD 662, and Stela 1 a decade later, on June 28, 672, perhaps in the reign of the same ruler. Centuries after the abandonment of Coba at the end of the Classic period, these and other stelae were venerated as sacred

---

* See Graham and Von Euw, 1997, for a compendium of images of Coba monuments, including excellent drawings of Stelae 1 and 5.

Stela 1 from Coba, Mexico. (Photograph by George Stuart)

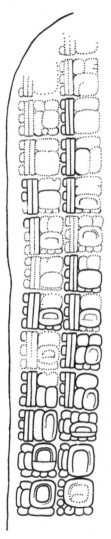

The lengthy Grand Long Count Date statement from Stela 1 at Coba. (Drawing by the author)

stones by Postclassic peoples in the fourteenth and fifteenth centuries, as indicated by the crude masonry platforms and walls built around the monuments. Carved out of a local porous limestone, the stones have not fared well out in the open for fifteen centuries. Each shows remains of relief sculpture on both sides, with images of standing rulers cradling large "ceremonial bars" in their arms and of ghostly captives at the rulers' feet.

Very long hieroglyphic inscriptions appear on both these Coba monuments, and each curiously mentions the familiar base date of the Long Count calendar, 13.0.0.0.0 4 Ahaw 8 Kumk'u. Here, however, the written dates take on a very strange appearance, involving far more than the standard five units of the Long Count. Instead, the mathematician-scribe opted to make use of *twenty-four units,* writing nineteen period glyphs *above* the bak'tun, each one bearing a number prefix of thirteen (two upright bars and three circles). One can see these in the long sequence of signs in the first two columns on the front of Stela 1. Many details are hard to see, but there's no question that this is a Long Count date, showing an absurdly long string of thirteens down to the point where we pick up the standard five units of the Long Count, written 13.0.0.0.0. So here we find 20 periods above and including the bak'tun bearing 13 as a coefficient. On the side of nearby Stela 5 we find the very same date with a lengthy string of 20 thirteens, then the last few units with "zero" coefficients. The stela is also eroded in many details, but just enough of it survives to show that we have two examples of the same date. Here is how we would transcribe them:

13.13.13.13.13.13.13.13.13.13.13.13.13.13.13.13.13.13.13.0.0.0.0
4 Ahaw 8 Kumk'u

It's interesting to see here yet another case of a juxtaposition between the numbers 20 and 13 (that is, 20 different units bearing the same 13 coefficient), in a way so different from their role as factors in the 260-day *tzolk'in*.

But what does it all mean? Why all these numbers?

It turns out that these lengthy sequences are representations of the Long Count calendar at full scale, with no shorthand omissions or abbreviations. The five periods that are so much more common and familiar—as in 13.0.0.0.0—actually are abbreviated glimpses of something far, far larger. In other words, the Coba dates show us the full picture of Maya time on a conceptual scale that is nearly incomprehensible.

Let's explore what this really means in terms of numerology and scale. For economy's sake, I will below abbreviate the just mentioned very long number using the "vinculum" symbol: the horizontal bar placed above a number to indicate repetition. In my new Maya usage, the bar goes above thirteen in the bak'tun's place to indicate that a whole series of thirteens repeat above it. For our purposes—and temporary sanity—the full Coba date can now be abbreviated in the following way, with the understanding that $\overline{13}$ stands for the twenty cycles above and including the bak'tun:

$\overline{13}$.0.0.0.0 4 Ahaw 8 Kumk'u

We know that a single bak'tun is equivalent to about 400 years (that is, 20 k'atuns), and that the k'atun below the bak'tun equals about 20 years (that is, 20 tuns). The same pattern works in the other direction, as we've seen, with each period growing exponentially out of the

The Grand Long Count

PERIOD	NUMBER OF DAYS	NUMBER OF TUNS (X360 DAYS)
+19	754,974,720,000,000,000,000,000,000,000	2,097,152,000,000,000,000,000,000,000
+18	37,748,736,000,000,000,000,000,000,000	104,857,600,000,000,000,000,000,000
+17	1,887,436,800,000,000,000,000,000,000	5,242,880,000,000,000,000,000,000
+16	94,371,840,000,000,000,000,000,000	262,144,000,000,000,000,000,000
+15	4,718,592,000,000,000,000,000,000	13,107,200,000,000,000,000,000
+14	235,929,600,000,000,000,000,000	655,360,000,000,000,000,000
+13	11,796,480,000,000,000,000,000	32,768,000,000,000,000,000
+12	589,824,000,000,000,000,000	1,638,400,000,000,000,000
+11	29,491,200,000,000,000,000	81,920,000,000,000,000
+10	1,474,560,000,000,000,000	4,096,000,000,000,000
+9	73,728,000,000,000,000	204,800,000,000,000
+8	3,686,400,000,000,000	10,240,000,000,000
+7	184,320,000,000,000	512,000,000,000
+6	9,216,000,000,000	25,600,000,000
+5	460,800,000,000	1,280,000,000
+4(Alawtun)	23,040,000,000	64,000,000
+3(K'inchiltun)	1,152,000,000	3,200,000
+2(Kalabtun)	57,600,000	160,000
+1(Piktun)	2,880,000	8,000
Bak'tun	144,000	400
K'atun	7,200	20
Tun	360	1
Winal	20	
K'in	1	

one that follows it, with the units read from left to right, from largest to smallest. So, above the bak'tun we have a unit that is made of 20 bak'tuns. The next highest in turn is equivalent to 20^2 bak'tuns, the next above that is 20^3 bak'tuns, and so forth, up to the alawtun.

Stelae 1 and 5 show that many larger periods exist above the alawtun, each exponentially larger than the last in the base-20 system. In this way we can simply do the math and see that the next higher period—it has no name as yet—represents 20 alawtuns, or about 1,280,000,000. Twenty of those make up the next, or about

25,600,000,000 years, and so on, and so on, and so on (See table, The Grand Long Count, on page 242). The very highest of these periods appears at the very top of the inscription, heavily eroded like the rest. It stands 19 units above the bak'tun, and therefore represents the following number of 360-day tuns or, roughly, years:

2,097,152,000,000,000,000,000,000,000 tuns

Now, this represents just *one* of these very highest units. Taking a look again at the Coba dates, we find that there are *thirteen* of these highest units indicated. That produces this number of tuns (again, roughly equivalent to years):

27,262,976,000,000,000,000,000,000,000 tuns

This isn't all. If we consider the full array of periods as shown on the two Coba monuments, computing 13 bak'tuns, 13 piktuns, 13 kinchiltuns, and so on, up to the very top, we are looking at an expression of a staggering amount of time:

$$13(20^{21}) + 13(20^{20}) + 13(20^{19}) + 13(20^{18}) + 13(20^{17}) + 13(20^{16}) +$$
$$13(20^{15}) + 13(20^{14}) + 13(20^{13}) + 13(20^{12}) + 13(20^{11}) + 13(20^{10}) +$$
$$13(20^{9}) + 13(20^{8}) + 13(20^{7}) + 13(20^{6}) + 13(20^{5}) + 13(20^{4})$$
$$+ 13(20^{3}) + 13(20^{2}) + 0(20 \text{ tuns}) + 0(\text{tuns}) + 0(20 \text{ days}) + 0(\text{days})$$

Which is the same as:

28,697,869,473,684,210,526,315,789,200 tuns

Now, recall that a single tun is 360 days long, just short of a solar year. So let's multiply the above number by 360 to see how many days are being expressed:

10,331,233,010,526,315,789,473,684,112,000 days

This, in turn, we can express as

28,285,978,483,664,581,446,157,328,238.631 years

Here, then, is the amount of elapsed time represented by the inscribed date on the Coba stelae, represented there in Maya fashion with the long sequences of thirteens. The Maya "Creation" of August 11, 3114 BC, took place more than twenty-eight octillion years after a *true* initial base date in the incomprehensible past. So, contrary to what some have often said, the 4 Ahaw 8 Kumk'u "era" date was not the absolute beginning point of the Long Count system. This really isn't too surprising, since some events of Maya mythology are said to have taken place well before 3114 BC, as we see, for example, in the inscribed tablets at Palenque and on the stelae of Quiriguá and Copan. It's therefore preferable to consider 13.0.0.0.0 4 Ahaw 8 Kumk'u as a relatively recent "Creation" among others, and a relatively late event within a far, far larger narrative about the gods and the cosmos. It's best considered as the base of the *current* bak'tun count, and a numerological station of great significance, when all of the units from the bak'tun and above were set at thirteen. The "true zero" date for the entire scheme, over twenty-eight octillion years ago, was the point at which all of the thirteens of these units were actually set back to zero. That implied date, while never written or directly referenced in a Maya text, would have looked like this:

0.0.0.0.0.0.0.0.0.0.0.0.0.0.0.0.0.0.0.0.0.0.0.0 4 Ahaw 18 Pax

We can perform the same exercise in the other direction, projecting forward to see how far the Grand Long Count will extend into the future. This is not quite so straightforward, however. (How can any of this be thought of as "straightforward," I imagine readers thinking.) Projecting ahead in time relies on a good understanding of the structure of the Grand Long Count, which has one strange quirk to it, as was hinted at in the previous chapter.

The 4 Ahaw 8 Kumk'u "base" of 3114 BC was formed out of a long
and steady accumulation of all of these periods, operating normally as
a vigesimal, or base-twenty, count, until all of the units reached thirteen. Then something weird happened. Normally one would expect
that $\overline{13}$.19.19.17.19 would turn on the next day into 14.0.0.0.0 3 Ahaw
13 Ch'en, but it didn't. Instead, the bak'tun number reverted back to
1, as 1.0.0.0.0.0 3 Ahaw 13 Ch'en. And we know that the same structure holds true for the higher piktun, which will change over from 13
piktuns to 1 piktun some 2,760 years from now. If two periods operate
this way, it's natural to infer that *all* of the periods of the Grand Long
Count undergo the same strange reset from 13 to 1, after which they
continue to accumulate days in the ways they always had.

Knowing now that the standard twenty units make up these future
periods, we can go about calculating the full capacity of the Grand Long
Count after the 4 Ahaw 8 Kumk'u base date. After that pivotal point in the
entire system, seven bak'tuns accumulate toward the turn of the next Piktun; seven Piktuns after that will lead to the end of the next k'inichiltun;
seven k'inchiltuns after that will end at the turn of the kalabtun, and so
on. In the end, the final day of the entire system will look like this:

> 19.19.19.19.19.19.19.19.19.19.19.19.19.19.19.19.19.19.19.19.17.
> 19 3 Kawak 17 Sotz'

Doing all of the math now gives us the size of the Grand Long Count,
spanning from 13.0.0.0.0 4 Ahaw 8 Kumk'u to its ultimate endpoint:

> 15,894,204,631,578,947,368,421,052,480,000 days =
> 44,150,568,421,052,631,578,947,368,000 tuns

which comes out to a bit over

> 43,517,152,096,098,311,708,523,306,538 years

Shave a little over five thousand years off this immense span of
time—the years elapsed between 3114 BC and now—and one has the

future capacity of the Maya calendar. So much for the Maya ever saying that the calendar would end anytime soon!

Now we can take this further and add this number to the elapsed span of time represented on the Coba stelae, starting at the absolute zero base date, when everything was set at zero. This sum represents the full amount of time that was *conceptually* represented by the entire system, incorporating past and future. Keep in mind that the Maya never wrote this; it's simply the capacity of the full calendar as I've described it:

26,225,437,642,105,263,157,894,736,592,000 days =
72,848,437,894,736,842,105,263,157,200 tuns =
71,803,130,579,762,893,154,680,634,776 years

The full Maya calendar therefore encompasses nearly seventy-two octillion years from beginning to end. Clearly, their conception of time will come to an end eons after our solar system has burnt itself out and the universe is no more.

At first the Grand Long Count seems to be a purely linear system, with smaller time units accumulating to make ever larger ones over an immense span of time. But looks are deceiving: it's not a straight linear system at all. For, integrated into this scheme are "rollbacks" at certain regular intervals. You'll recall that the first bak'tun after 13.0.0.0.0 was 1.0.0.0.0—*not* 14.0.0.0.0, as one might at first suppose. In other words, the bak'tun sequence at this particular moment was cut short of its normal course, not allowed to progress beyond thirteen. Once this strange shift in time took place, it did not necessitate a change in the piktun, the next highest period. In this way piktun thirteen was exceedingly long, composed of thirty-three bak'tuns instead of the normal twenty bak'tuns. Perhaps I can illustrate this difficult concept in the following way, writing a sequence of individual bak'tuns. Notice that the piktun number (the sixth from the right) is the same over the course of so many bak'tuns, each separated by about four hundred years. Only two Long Counts repeat, shown in boldface.

1̄3̄ .13.13.0.0.0.0	4 Ahaw 8 Kumk'u	August 11, 3114 BC
1̄3̄ .13.1.0.0.0.0	3 Ahaw 13 Ch'en	November 13, 2720 BC
1̄3̄ .13.2.0.0.0.0	2 Ahaw 3 Wayeb	February 16, 2325 BC
1̄3̄ .13.3.0.0.0.0	1 Ahaw 8 Yax	May 21, 1931 BC
1̄3̄ .13.4.0.0.0.0	13 Ahaw 13 Pop	August 23, 1537 BC
1̄3̄ .13.5.0.0.0.0	12 Ahaw 3 Sak	November 26, 1143 BC
1̄3̄ .13.6.0.0.0.0	11 Ahaw 8 Woh	February 28, 748 BC
1̄3̄ .13.7.0.0.0.0	10 Ahaw 18 Sak	June 3, 354 BC
1̄3̄ .13.8.0.0.0.0	9 Ahaw 3 Zip	September 5, AD 41
1̄3̄ .13.9.0.0.0.0	8 Ahaw 13 Yax	December 9, 435
1̄3̄ .13.10.0.0.0.0	7 Ahaw 18 Zip	March 13, 830
1̄3̄ .13.11.0.0.0.0	6 Ahaw 8 Mak	June 15, 1224
1̄3̄ .13.12.0.0.0.0	5 Ahaw 13 Sotz'	September 18, 1618
1̄3̄ .13.13.0.0.0.0	**4 Ahaw 3 K'ank'in**	**December 21, 2012**
1̄3̄ .13.14.0.0.0.0	3 Ahaw 8 Tzek	March 26, 2407
1̄3̄ .13.15.0.0.0.0	2 Ahaw 18 K'ank'in	June 28, 2801
1̄3̄ .13.16.0.0.0.0	1 Ahaw 8 Xul	October 1, 3195
1̄3̄ .13.17.0.0.0.0	13 Ahaw 18 Muwan	January 3, 3590
1̄3̄ .13.18.0.0.0.0	12 Ahaw 18 Xul	April 7, 3984
1̄3̄ .13.19.0.0.0.0	11 Ahaw 8 Pax	July 11, 4378
1̄3̄ .1.0.0.0.0.0	10 Ahaw 13 Yaxk'in	October 13, 4772

As shown here in the last line, the shift from thirteen to one in the bak'tun did not necessitate any change in the piktun. The piktun's number—to the left of the bak'tun—remains at thirteen throughout. But this is not a static number; it, too, must shift, given the scale of Maya time computations into the distant past and future.

One extremely important text from Palenque tells of a *future* mythological event that will occur on the end of a piktun on 10 Ahaw 13 Yaxk'in, on the day October 13, 4772. According to this record, this will be the end of *one* piktun, after a full sequence of nineteen bak'tuns have passed. This tells us something very important: that, like the bak'tun change in 3114 BC, the Piktun number will shift from thirteen to one, *not* to fourteen. The inescapable implication is that *each of the thirteens shown in the Grand Long Count will revert to one and then prog-*

A key passage from the Temple of the Inscriptions at Palenque, describing the upcoming turn of the Piktun period as "1 Piktun" (at the middle of the right column). (Drawing by the author)

ress once again as normal up to nineteen before requiring the next higher period to shift forward. For this reason the coefficient thirteen will automatically appear *twice* within the span of the oddly extended period above it. And because the number coefficients on the higher periods reverted to one after thirteen, there are several similar "kinks" in the system that are at once maddening and fascinating, where complete sequences of thirteens mathematically recur. In this respect the system isn't completely linear in its structure. It may have a linear façade, but it incorporates several feedback loops where time can repeat itself. In this respect it's very Maya.

Lest the reader be confused beyond all hope, let's illustrate the structure in a slightly different way. The Creation base date of 3114 BC *does* represent a point in a purely linear system of accumulated time, where all periods above and including the tun represent a base-twenty system of place notation.

13.13.13.13.13.13.13.13.13.13.13.13.13.13.13.13.13.13.13.0.0.0.0
4 Ahaw 8 Kumk'u

No hiccups or backward loops have occurred in the system up to this point. As we've seen, this equals a little over twenty-eight octillion years from what we can call the zero base date, the true starting point of everything. With the passing of thirteen more bak'tuns, we come to December 21, 2012, and its corresponding position in the Grand Long Count *is identical,* with the only change found in the month position of the Calendar Round:

13.13.13.13.13.13.13.13.13.13.13.13.13.13.13.13.13.13.13.0.0.0.0
4 Ahaw 3 K'ank'in

Because *two* occurences of thirteen bak'tuns exist within thirteen piktuns, a complete repetition of the Grand Long Count date is inevitable. So, the 2012 date of the Maya calendar is not actually the "end" of anything; rather, it's a *mathematically predetermined recurrence of the date of ancient Maya Creation.* To reiterate, there will be a number of such repetitions in the distant future as well, because once thirteen piktuns reverts to one piktun, the process begins anew. Twelve piktuns after that, we have a repetition of thirteen piktuns, which inevitably must include another thirteen bak'tun station. So, about a hundred and ten thousand years from now, our descendants will encounter a date that looks much like the others we've already seen:

13.13.13.13.13.13.13.13.13.13.13.13.13.13.13.13.13.13.13.0.0.0.0
10 Ahaw 13 Ch'en

As time passes, the same mechanisms hold true for the periods above the Piktun. This way 2,310,370 years later, another repetition takes place:

13.13.13.13.13.13.13.13.13.13.13.13.13.13.13.13.13.13.13.0.0.0.0
13 Ahaw 3 K'ayab

The full system can accommodate twenty such repetitions, corresponding to the twenty periods above and including the bak'tun. In this way the 2012 event, far from being the end of the Maya calendar, is the first of these many built-in Long Count recurrences of Creation day. Many have made note of the fact that the 4 Ahaw 3 K'ank'in date in 2012 falls on or near winter solstice. Assuming that our correlation is about right, this is indeed the case. But I find it interesting that the many future repetitions of the "many thirteens" don't fall on astronomically important dates. Could the winter solstice connection be a coincidence, then? Perhaps, but I am willing to entertain that Maya astronomers were very happy with the mathematical result, fortuitous or not.

So the full Grand Long Count, from its absolute base point to its full capacity in the remote future, was full of repeating permutations involving a day of profound cosmological and religious importance, when "the gods were set in order." Its staggering magnitude rests on a numerical contrivance built on the interplay of the sacred numbers of thirteen and twenty as well as the symbolism of the day 4 Ahaw. The full calendar's beginning point, well before the Big Bang of our own universal cosmology, mathematically began the day 4 Ahaw (if there had been a sun and an earth to make a day possible, that is). The establishment of the gods and the universal order for the Maya took place also on 4 Ahaw, in the year 3114 BC. At the end of the entire calendar, once all of the periods have been played out and far after our world is gone, 4 Ahaw will occur again on the anniversary of this same date.

Cosmic History

Let's bring this down to earth if we can. Putting aside for the moment the awesome scale of the Grand Long Count, we should consider anew the measly five periods that most of us are used to using in calculating Maya dates—the baktun, k'atun, tun, winal, and k'in. It should be clear by now that every conventional Long Count date using these five units is an abbreviation of this far larger structure. In other words, all Maya dates can be represented in this lengthy scheme even though practical matters of space and economy naturally kept scribes from needing to include the high periods, leaving all those repeating thirteens on the cutting room floor. Most historical and mythical records seldom ever required anything above the four-hundred-year bak'tun to provide meaningful chronological perspective, so the five most basic parts of the Long Count sufficed to show time's linear progression over the course of centuries and millennia.

One inscribed block at Yaxchilan offers proof that regular Long Count dates could be written in this expanded format, if the context so demanded. The block comes from a intricately carved step in front of the site's largest temple, Structure 33, where the carving on each stone block shows an individual member of Yaxchilan's royal family engaged in a "reenactment" of a ball game said to have taken place in the deep mythic past—an illustration of how real ceremonies were so often based on the notion of the kings and nobles as gods and heroes, and how, through performance, they brought obscure religious narratives to life. In the inscription above one of these scenes we find a series of glyphs very reminiscent of the huge Long Count dates from Coba, including a series of repeating thirteens. There are not so many here, however—just eight periods above the bak'tun. If we look at the final five numbers, they look for all the world like a regular Long Count recording a date in historical Maya time, 9.15.13.6.9, followed by the day and month positions, 3 Muluk 17 Mak. We can transcribe the date this way:

13.13.13.13.13.13.13.13.9.15.13.6.9 3 Muluk 17 Mak

There's little doubt that this is simply an extended way of writing the historical date 9.15.13.6.9 3 Muluk 17 Mak, when, according to the rest

A Long Count date from Yaxchilan, showing higher periods above the Bak'tun set at 13. (Drawing by the author)

of this text, the local ruler Bird Jaguar dedicated the steps of Structure 33. As part of that ritual, he also performed a mock ball game that saw the sacrifice of a prisoner named "the Guardian of Black Deer, the Lakamtuun Lord." This unfortunate man is depicted in the scene as a bound figure inside the figurative ball rolling down the temple steps. Writing the date in such a purposeful way brought this event in history into the mythological mix, as it were. Even without all of the thirteens, any date in Maya history could be understood to have its proper place and orientation in the grander scheme of things, fitting neatly within the calendar encompassing almost seventy-two octillion years.

The Maya also recorded many "pre-Creation" dates that emphasize the by now obvious point that the base date of 3114 BC was in no way the starting point of their calendar systems. These mythical episodes indicate that well before all the gods were arranged in their proper order, and well before the three stones of the Creation hearth were dedicated, gods and heroes were active in primordial time, a period of ancient "prehistory."

In the three famous temples of the Group of the Cross at Palenque, for example, one tablet opens with the Long Count date 12.19.13.4.0 8 Ahaw 18 Tzek. The bak'tun unit is set at twelve, much higher than what we'd expect in a conventional historical date, where the bak'tuns are set at eight, nine, or ten, according to the span of recorded Maya history. We might therefore be reasonable in thinking this is a projection far into the future for the Maya scribe who composed it, equal to April 19, 2006. But the Calendar Round is not what we would expect, either. Instead, this is a pre-Creation date, referring to the bak'tun *before* 3114 BC. Recall that the Creation base date was 13.0.0.0.0 4 Ahaw 8 Kumk'u, implying of course that any date before that would have a bak'tun with a value of twelve. The Palenque date comes some seven and a half years (twenty-eight hundred days) before Creation, and corresponds to the mythical birth of a major Creator deity who engendered the three patron gods of the site's local dynasty, known as the Palenque Triad. As it's written (without our doing the math), the distant mythical date looks much like a historical one, a mirroring of myth time and real time. This was no doubt intentional.*

Another Palenque tablet, excavated not long ago from a tree-covered mound in the forest surrounding the site, follows much the same pattern and makes even more overt links between myth and history. Temple XIX was a shrine built late in Palenque's history, in AD 734, during the reign of a king named K'inich Ahkal Mo' Nahb. He was likely the grandson of the famous ruler Pakal, and nephew of K'inich Kan Bahlam, who oversaw the dedication of the three principal temples of the Group of the Cross. When excavated, Temple XIX was found to enclose an unusual small platform or pedestal inside its walls, tucked into a corner where it must have been difficult to see when the temple was built. Two sides of the platform were decorated with stunningly beautiful carvings in low relief, one showing members of Palenque's royal court surrounding K'inich Ahkal Mo' Nahb on the day of his accession.

A long text uniting themes of myth and history accompanies the scene, opening with one more odd-looking Long Count: 12.10.1.13.2 9 Ik' 5 Mol. This is, once again, a pre-Creation date, corresponding to March 8, 3309 BC, when, reading on in the text, we find that the god GI, a solar deity who was the principal member of the Palenque Triad of gods, was "seated in the kingship . . . in the resplendent sky"—evidently there were kings before Creation, though it's difficult to say what or whom they ruled over. Much later in time, in 2325 BC, we read about the accession to office of the Maya maize god, who earlier had engendered the Palenque Triad gods through his own bloodletting and sacrifice in the post-Creation world. This took place on the date 2.0.0.10.2 9 Ik' Seating of Sak. And later still we read about the accession to office of the Palenque king K'inich Ahkal Mo' Nahb, on the day 9.14.10.4.2 9 Ik' 5 K'ayab, or December 28, 721. Now, look closely to see what might link these three dates

---

* See D. Stuart and G. Stuart, 2008, for an overview of Palenque's history and mythology. A more detailed treatment of the mythical context of the inscriptions from these temples can be found in Stuart, 2006.

Accession scene of the Palenque ruler K'inich Ahkal Mo' Nahb, from a relief tablet discovered in Temple 19. (Drawing by the author)

spanning four thousand years. You'll notice that the dates look quite different save for their common position in the 260-day round. This establishes that the accession of the maize god in 2325 BC and the historical crowning of Palenque's king in AD 721 are both *tzolk'in* (260-day) anniversaries of the foundation seating of the god GI, well before Creation ever occurred. Even the artists of the Temple XIX platform were keen to show the parallel, by showing K'inich Ahkal Mo' Nahb impersonating GI on the day of his accession, as if he were a reincarnation of that very ancient manifestation of the sun. To the Maya, history was often about tracing similar strands and connections between the present time and mythic time, showing how they serve as reflections of each other.

The most outrageous calculations into the deep, deep past come from inscriptions on ruins of Quiriguá, the same place where many details of the Maya Creation event were recorded on Stela C.[2] A number of the monuments there make reference to ancestral gods and rulers through a similar parallelism in narrative structure, where events of the past and the present are connected and juxtaposed over truly vast scales of time. One text on a monument today called Stela A mentions the completion of nineteen high periods from the Grand Long Count, probably the unit that comes eight places *above* the bak'tun. Now, one of these units equals a little over ten trillion years. On Stela A, there is record of nineteen of those same units "ending" in the distant past. I haven't made the calculations to check if this is correct, but the scale of time we're talking about would probably be about 200 trillion years. And what happened on that date? According to the text, some mythical persona "ended" the period and basically celebrated a special period ending, just as the king K'ahk' Tiliw Chan Yopat did in contemporaneous history. Numerous other stelae at Quiriguá make similar claims, projecting varying lengths back in time to link the king's ceremonies to similar ones performed eons before.

Obviously from these records we see that the Maya were keenly interested in bridging myth time and history, and to do so they continually sought to find and document symmetries between mythic dates and events in contemporary experience. It's quite possible that they even fudged the data from time to time to make the symmetries they so loved to display in their sacred texts.

The deep time of the full Maya calendar is stunning in its scale and in the virtuosity displayed by its internal mechanisms. I think it's fair to say that it constituted the grandest expression of time ever put down on stone or paper by any human mind. It certainly dwarfs our own understanding of the vast temporal scale of the universe. In some way we can look on it as a purely numerological contrivance: an exponential count involving twenty units above the k'atun, all multiplied by thirteen. I doubt they could truly grasp the scale of that any better than we could, for it was indeed a kind of numbers game they were playing. It demonstrated a virtuosity that I can only admire and try to understand in its full extent. But it was a serious game, too. The Grand Long Count produced a numerological symmetry that I find stunning in its beauty, and expressed a time scale that I suspect not many mathematicians could even attempt to fathom before the days of the computer.

In the Mesoamerican framework of time, humans and our history occupy a truly minuscule place—just a handful of bak'tuns at most. But there was so much more going in the vastness of Maya time and space than just human events, with their accessions, kingly ceremonies, wars, and royal marriages. As the sacred inscriptions of Palenque, Quiriguá, and other places make clear, the far distant past and the far distant future were, and will be, the time of gods.

9

KINGS OF TIME

I am the substance of heaven, the substance of clouds.
—Itzámat Ul, the deified king of Izamal,
Yucatán, on being asked who he was.[1]

Domingo Falcón looked over the huge flat stone and saw that it was carved. He had found it while clearing brush from an ancient mound near his house in Coba one warm afternoon in the summer of 1975, the same summer the village experienced its awful drought. Word of Domingo's find spread quickly and soon reached my father as he worked inside our rustic Maya house, built near the lake among the remains of many more of Coba's collapsed and ruined temples. I remember that day well, especially hearing the excitement in my dad's voice as we all piled into our Chevy Suburban to go check out the new discovery. A few minutes later we were gathered on the slope of the mound looking at the large fragment, obviously a piece of something much bigger, covered in faintly preserved hieroglyphs. My father read a Long Count date out loud: "9 . . . 17 . . . 10 . . . 0 . . . 0." Back at our hut, after a little computation and checking of published tables, he told us that this was November 26, AD 780. The broken slab turned out to be the upper piece of a large stela that was lying on the ground only a few meters away, still preserving its solemn portrait of a powerful Maya king holding his ceremonial staff and surrounded by bound prisoners. The top piece had been missing for years, and now, suddenly and unexpectedly, the stela was finally complete. I'm pretty sure that from that hot, dusty, thrilling moment I was forever hooked on Maya archaeology.

Stela 20 is not the prettiest of Maya carvings—Coba's sculptors lacked a certain flair we find in other Maya kingdoms—but it's still an impressive monument, and clearly dedicated to the story of a very important man. It's also a good example of the type of monument known as a stela, a term borrowed from Greek archaeology: basically an upright slab set into the earth, either outside or inside a shrine, bearing a carved scene or text. Such monument types are found all over the world: in Mesopotamia, Egypt, and China. Stelae may seem exotic to us, but we've used them, too, in our own recent history—for example, the elaborately decorated seventeenth- and eighteenth-century gravestones still standing near my old house in Marblehead, Massachusetts, are basically the same idea. Large stelae appear throughout ancient Maya ruins, both large and small, most dating to the Classic period, before about AD 850. They were used to mark stations of the Long Count calendar, and specifically to commemorate a ruler's ceremonial duties on those auspicious days. Coba's Stela 20 was dedicated on 9.17.10.0.0 12 Ahaw 8 Pax, at the halfway point of a k'atun, and the ruler depicted on the stone—we don't know his name—probably commissioned others at intervals of five tuns.

The imagery of stelae can vary widely from region to region, but they typically feature a standing portrait of a king or queen in his or her powerful glory, holding symbols of office and cosmological power. The small bound captives shown at the feet of the Coba ruler are obvious symbols of military might, and they reveal how artisans could use such monuments to convey a number of related messages about royal authority. On Stela 20, as in many other examples, a king appears to his audience as not just a victorious warrior but also a cosmic actor, playing a direct role in perpetuating time and seeing to its proper care and passage. The prisoners at his feet may well have been participants in the ritual, probably as sacrificial victims. When my father reunited the pieces of Stela 20 more than thirty years ago, he reconstructed what amounted to an ancient billboard touting one Maya king's engagement with sacred time in AD 780, and revived a sacred stone with its varied overlapping messages about ideology and kingship.

Kings and the mechanisms of sacred time were reliant upon one another in a way that seems strange to Western concepts of authority and divine rule. On the one hand, time periods, as animate beings in some sense, depended on kings to oversee their proper development and growth. The ritual texts of the Classic period make frequent reference to kings overseeing the completion of time periods as if they were tending to a cornfield, and "harvesting" time once the cycle of growth was complete. Rulers also had the duty to mark time's passage through the erection of stone monuments, many of which still dot the ruins of Maya cities. On the other hand, this basic duty of kingship allowed rulers to demonstrate the sacred underpinnings of their royal office. As far as the Long Count is concerned, I don't think it much of an exaggeration to say that for the ancient Maya, the calendar was kingship, and kingship was the calendar. This fundamental relationship played itself out in a number of ways, including, I argue, having profound effects on the course of ancient Maya history itself.

The basic title for Maya kings was *k'uhul ajaw*, "holy lord."* The adjective *k'uhul* derives from the noun *k'uh*, meaning "god," and emphasizes the fundamental divinity of kings, as opposed to other nobles and lords, who would have existed at any of the numerous royal courts of the Classic period. They were the most prominent and divinely connected members of a larger class of nobles who went by the more generic honorific term *ajaw*, "noble" or "lord"; *ix ajaw*, "noblewoman," was its female form. The word probably conveys an idea of mastery or

---

* The *k'uhul ajaw* royal title was first identified by the author in the late 1980s. In the hieroglyphs, this title is most often seen as the basic component of so-called emblem glyphs, local titles that label royal seats of authority. An emblem glyph at Palenque would read *K'uhul Baakal Ajaw* (Holy Baakal Lord), and one from Kalalmul would read as *K'uhul Kan Ajaw* (Holy Snake Lord), for example. See Stuart, 1993.

ownership over others, as indicated by its meaning when used in some present-day Mayan languages. Many of the elite members in classic Maya society were *ajawoob,* but possibly not all. I suspect that the term implies a role of ownership and rule over a certain segment of society and also over property, a role incorporated in the rigid but still poorly understood hierarchy of Maya economics and society.

If we were to transport ourselves back to some ancient city such as Calakmul or Palenque, we would have little trouble recognizing an *ajaw.* I suspect he would not be walking about the plazas or markets. Instead, they spent much of their time in fairly enclosed and restricted spaces, in palaces of various sizes, conducting official and ritual business from their throne rooms. At times they might be seen transported about the city in elaborate litters or palanquins borne by slaves or workers. The accoutrements of their high status—jade jewelry, fine cloth, and (in ceremonies) quetzal plumes—would have set them apart from lower classes of people in the community.

Maya rulers were more than ceremonial figureheads. Like their Aztec counterparts, they probably oversaw a great deal of the practical matters of ruling their kingdoms, acting as judges, planning war, engaging in diplomacy, and setting economic rules and policies. Much of this is supposition, however, since virtually nothing is known of such daily aspects of kings' work and duties; we have to infer from what we know from later historical sources and extrapolate backward. Perhaps because of the silence of ancient texts on such mundane nonritual aspects of rule, it's sometimes said that Maya kings were notoriously disengaged from the real issues surrounding society as a whole. One bestselling book touching on the ancient Maya collapse even goes so far as to say that Maya kings were "too self-absorbed in their own pursuit of power to attend to their society's underlying problems."[2] This seems a bit unfair, especially given how little we know of Maya systems of governing. The Maya collapse itself was certainly a complex phenomenon that involved, as we will see, significant awareness of temporal cycles and their perceived influence on human affairs.

What we do know about Maya kings and other Mesoamerican rulers tells us that they were extremely conscious of their roles as active caretakers of time's regular progress. As divine beings they were perhaps even considered members of the wider class of "heavenly gods and earthly gods" who pervaded the cosmos, representing the interwoven aspects of time as animate beings and as forces of nature. In this chapter we will explore these complex ideas in more detail, showing that for the ancient Maya, the understanding of time was inseparable from the roles and duties of kings.

Kings as Periods, Periods as Kings

To begin to grasp this connection between rulers and time periods, it's important to recall that all period endings occur on the same day of the 260-cycle, Ahaw. Any tun, k'atun, or bak'tun ending, up to the highest period of the Grand Long Count, fell on this single day sign. The name of the day in Yukatek, *Ahaw,* is the same word we find in the ancient texts as a basic title for rulers, *ajaw,* "lord, noble." The symbolism of this association must have been intentional, since these days were, like the year bearers of central Mexico, the "rulers" of the period. And by extension this idea came also to involve the kings themselves, the *k'uhul ajawoob,* who upon assuming the throne likewise took on significance as the rulers or patrons of tuns, k'atuns, and other time periods.

Artists who designed inscriptions and scenes of rulers some-

A Maya king (ajaw) *as the day sign "9 Ajaw," from a stone altar in the Museo Nacional de Arqueologia e Etnologia de Guatemala, Guatemala City.* (Drawing by the author)

times used visual playfulness to convey this key overlapping of time and the royal person. A number of monuments that feature period-ending dates show very prominent *Ahaw* day signs with a number prefix—what we simply like to call "giant *Ahaw* altars." One such monument, now in the National Museum in Guatemala City—since it was looted, its place of origin is unknown—shows a large 9 Ahaw glyph in the center of a text. The date commemorated by the stone is 9.15.15.0.0 9 Ahaw 18 Xul (May 29, AD 746), and the inscription records the various rituals the king performed on that day, including the casting of incense. Hieroglyphs dominate the composition, but within the enlarged day sign we see a portrait of the historical king as the day sign *Ahaw*, "king." His name is difficult to read in the accompanying inscription, but his title is "Guardian of the Nine Stone." I suspect this is a reference to the altar itself, named "Nine Stone" because of its commemoration of the time period ending on the day 9 Ahaw. A stela from another site called El Palma shows the same thing: a giant *Ahaw* glyph that is at once the initial element of the inscription and a portrait of the local king who ruled over an ancient polity called Lakamtuun. Inside the cartouche of the day sign, the real *ajaw*, "lord," holds a ceremonial object, probably a ritual knife used in bloodletting ceremonies associated with the period ending.

The k'atun periods themselves were considered rulers who were enthroned every twenty years, depicted as animate characters, a series of individual lords who sit upon cosmic thrones. In the Paris Codex (a Late Postclassic Maya book like the Dresden and Madrid codices), thirteen sequential pages, referred to as the k'atun pages, highlight several such scenes, where different k'atuns appear as enthroned gods on elaborate seats, each made of a sky band and a cosmic alligator with a drooping head. These are, no doubt, the "idols" of the thirteen k'atuns described in early sources from Yucatán, and also by Avendaño in his treatise on Itzá religion. In each of the larger scenes, another god stands before the cosmic seat, presenting offerings, and among the complex array of elements we see labels such as "13 Ahaw," "11 Ahaw," and so forth, naming each period. The poor preservation of these pages makes their study extremely difficult, but we can definitely link them with similar scenes found on earlier stone monuments from the Classic period. A small, beautifully incised bone now in the Dallas Art Museum shows precisely the same scene as the Paris Codex k'atun pages, although here the date has nothing to do with k'atun periods, at least not directly. Rather, the scene on the bone is one of crowning and accession to power, involving a young god seated in a similarly elaborate bench or throne. (He may be the maize god, although identification is not completely certain.) But more interesting and revealing parallels can be found at Piedras Negras, Guatemala, in the famous "ascension" stelae that Tatiana Proskouriakoff studied in her breakthrough paper on the existence of dynastic history on Maya monuments. These consistent scenes again show a person perched atop a high celestial seat, built up almost as if upon a small scaffold. But these are not the gods we see in the other scenes. These are historical kings of Piedras Negras, shown in their first official royal portraits. Proskouriakoff and others interpreted them as actual scenes of accession and enthronement, but in fact they are highly metaphorical, linking the crowning of the king with the coming of a new k'atun or other period. Each

Stela 6 from Piedras Negras, depicting a new ruler of the local dynasty on a cosmic calendar throne. (Photograph by the author)

A misconstrued "k'atun wheel" from the seventeenth century, depicting thirteen deceased Maya rulers of Yucatan. Its original form was surely a "k'atun wheel" representing a sequence of thirteen k'atun dates.

of these "ascension" monuments was dedicated on the first period ending of a new king—not on his actual crowning date—so they serve as artistic fusions of history and time. The new *ajaw* of Piedras Negras merges his identity with the new *Ahaw* date of the period ending, assuming the role of that k'atun or other time period.

A so-called k'atun wheel from the Book of Chilam Balam of Chumayel, *depicting a cycle of thirteen k'atun periods.*

To me this playful interaction between the identities of kings and time periods was a hallmark of Classic-period art and royal ideology, though elements of it can be seen still in operation much, much later, well into the colonial period. The best survival I know is the very European-looking "coat of arms" illustrated by the historian López de Cogolludo in his *Historia de Yucatán,* published in 1688. The form of this shield or crest is almost certainly a European misreading of a traditional Maya "k'atun wheel," as Daniel Brinton recognized over a century ago. The image depicts the thirteen Maya men, each named by a caption, who, according to Cogolludo's account, were all lords of the town of Mani, all of whom died in a massacre planned by their rival faction, the Cocom.* Each head is shown with its eyes closed in death. But Cogolludo misunderstood the image he copied in the town of Mani. (The original is now lost.) For if we compare these images and the names to certain parts of the native Books of Chilam Balam, we find that this coat of arms is in essence a "k'atun wheel," a depiction of the thirteen k'atuns of a 260-year cycle. In fact, the very same names are found in two of

---

* The massacre of the Mani lords during their pilgrimage to Chichen Itzá in 1536 was a defining event in the history of Yucatán in the early sixteenth century. It was at heart a violent episode in a protracted conflict between two ruling lineages of Yucatán at the time, the Xiu and the Cocom, and it set the stage for years of continued strife in the region. For a summary of relevant sources on the episode, see Restall, 1998, p. 204.

the Chilam Balam documents (those of Mani and Kaua), where they are used as the designations of individual k'atun periods in history. For example, in the Kaua manuscript we read the following tally of sequential k'atun eras:

Katun 3 Ahau.* *Ah Napot Xiu* is its name.
Katun 1 Ahau. *Zonceh* is its name.
Katun 12 Ahau. *Ahau tuyu* is its name.
Katun 10 Ahau. *Xul cum chem* is its name.
Katun 8 Ahau. *Tu cuch* is its name.
Katun 6 Ahau. *Cit Couat Chumayel* is its name.
Katun 4 Ahau. *Uluac chan* is its name.
Katun 2 Ahau. *Nauat* is its name.
Katun 13 Ahau. *Ah Kinchy Coba* is its name.
Katun 11 Ahau. *Yiban Caan* is its name.
Katun 9 Ahau. *Pacaab* is its name.
Katun 7 Ahau. *Kan Cabaa* is its name.
Katun 5 Ahau. *Kupul* is its name.†

This list provides fascinating evidence that the twenty-year k'atun periods were each named for a particular individual in history, all perhaps rulers of Yucatán's various provinces.‡ It's difficult to know

* *Katun* is an older spelling of *k'atun;* similarly *Ahau* is a traditional spelling of the Yukatek day name we write here as *Ahaw*.

† This list is adapted from Morley, 1920, p. 482. Morley was, as far as I'm aware, the first to see through Cogolludo's mistake, and note that the coat of arms was a Europeanized misreading of a katun wheel.

‡ The nature of this interaction between historical kings and time periods is very difficult to tease out from the sometimes contradictory evidence found in the various Books of Chilam Balam. The ruler Ah Napot Xiu, for example, is said to have died a century before the period attached to his name, katun 3 Ajaw. It could well be that a list of various kings and lords of Mani were "grafted" onto the 260-year K'atun wheel by later native historians in the colonial period, who wished to emphasize the structure and correspondence between periods and lords over ancient historical veracity.

just who all of these named characters are, but I suspect they may have been lords or priests of distinct provinces, either contemporaneous with one another or from different eras. We should recall the direct statement by our old friend Fray Avendaño y Loyola, who described the political geography of pre-Conquest Yucatán as a reflection of this conceptual melding of time, space, and individual person:

> (The) thirteen ages are divided into thirteen parts, which divide this kingdom of Yucathan, and each age, with its idol, priest, and prophecy, rules in one of these thirteen parts of this land, according as they have divided it; I do not give the names of the idols, priests or parts of the land, so as not to cause trouble, although I have made a treatise on these old accounts with all their differences and explanations, so they may be evident to all, and the curious may learn them, for, if we do not understand them, I affirm the Indians can betray us face to face.[*]

Thus, like the kings of Piedras Negras depicted on their sky thrones as "new *ajaws*," the rulers who governed in Yucatán shortly before the arrival of the Spanish may well have fused their identities with those of the rulers of the major time periods in history. For the Maya, all politics were far from local. They were cosmic.

By the Late Classic era, the Long Count calendar was a political tool and artifact, a concious reflection of the rulers' desire to be seen somehow as embodiments of cosmic time and its recurrent cycles. No wonder, then, that the celebration of the Long Count calendar on stone monuments essentially died out alongside the demise of Classic Maya ideology and kingship. After AD 900, rulers severed their ties to the notion that the Long Count system was a reflection of their own grandeur and responsibility as keepers and tenders of the cycles.

---

[*] See Avendaño y Loyola, 1987, p. 39. His mention of his "treatise" explaining the details of the katuns and their meanings is tantalizing; no copy has ever been found.

Time Rites

Most freestanding monuments found at Maya ruins were erected in order to commemorate period endings and the local ruler's associated ritual performances. But just what were these rituals and ceremonies? It's sometimes not too easy to know just what Maya kings were doing on these very special occasions, even on a superficial level, and one can be sure that the rituals they did perform were deeply invested with rich cosmological meaning and symbolism. Take, for example, the scene

Altar from El Cayo, Mexico, depicting the casting of incense on a k'atun ending by a local ruler. (Drawing by P. Mathews)

depicted on a disc-shaped altar excavated at the site of El Cayo, Mexico, by my colleague Peter Mathews. It bears a beautifully naturalistic portrait of El Cayo's local ruler Aj Chak Wayib, a masterwork of Maya portraiture. The carved date on the side of the stone provides a Long Count of 9.15.0.0.0 4 Ahaw 13 Yax (August 16, 731), a major period ending of the Late Classic period, when the bak'tun was three quarters complete. In the carved scene of his ceremony on that day, Aj Chak Wayib sits before a small stone table that supports an elaborate ceramic brazier. The table is perhaps the carved altar itself (a typical Maya use of self-reference in art), and upon it he casts a number of small pellets of incense from his outstretched hand. The caption above the scene states that the ceremony is called *k'altuun*, "stone-binding," taking place "when 15 k'atuns end" or are "re-planted."

What does this mean, precisely? First of all, it's important to point out that the similarity between the words *k'atun* and *k'altuun* is not coincidental. The word used in ancient Yucatán for the twenty-year period derives literally from *k'altun*, "twenty tuns," which is in turn related to the more archaic phrase *k'altuun*, "stone binding," which we see in the hieroglyphs. (The Yukatek word for "twenty" derives from the verb for "to enclose, fasten." Remember, too, that the use of the word *k'atun* for the calendar period was probably a late innovation, not used by the Classic Maya.) "Stone binding" was a direct description of one of the most important calendar ceremonies from the Classic period, when sacred stones that symbolized the individual time periods were ritually bound or wrapped. It was one of the ways, also, that Creation was described on a stela from Quiriguá, where we read of three stones being bound or collected together in 3114 BC. We find these wrappings represented on occasion in artistic imagery, probably made of the same sacred bark paper used for the head scarves of Maya kings. In a sense, the stones are inaugurated much like an *ajaw* on the day of his inauguration.

The so-called scattering ritual depicted on the El Cayo altar and on many stelae was also tremendously important. Texts mention that this involved the casting of a certain type of small incense or resin called *ch'aah*, and usually this would have been thrown into a large brazier or

Monuments commemorating different period endings of the Long Count, today standing in the main plaza of Copan, Honduras. (Photograph by the author)

container of a ritual fire. In other cases it may refer to the sprinkling of blood droplets as offerings in self-sacrifice by the king; there's often ambiguity about the nature of the rite when no image accompanies a text. The casting of incense on stone tables and other surfaces resembles the techniques of calendrical divination found up to the present day in Mesoamerica. It's entirely possible that in performing their binding and scattering rites, Maya kings were participating in a tradition of calendar ritualism that in part survives up to the present day.

Scenes of scattering rites or other period-ending ceremonies typically appear on stelae erected in plazas or in front of temples. The size and dimensions of some of these monuments suggest they may have been considered as stand-in "bodies" of the performing king, living portraits engaged in public ceremonies out in the open. The well-preserved monuments that still stand in the plaza of Copan offer a good case in point. Here the images of the ruler were erected on many period endings over the course of two decades, accumulating over time to the point where the multiple statue-like portraits resemble an ancient diorama. All of them show the Copan king Waxaklahuun Ubaah K'awiil in different ritual guises at different points during his reign. By the end of his rule (he was captured in war by the king of Quiriguá), the plaza had become a place where one could see his images "frozen" in time, each engaged in a perpetual ritual performance. Many monuments at Maya sites operated on this principle, I think, where artistic images and royal portraits conveyed more than mere representation; they were themselves animate embodiments of the king, extensions of the kingly self that always "acted" to ensure the perpetual renewal of time and the cosmos.

This important notion that kings were world renewers resonates with what we've already learned about them and their ritual duties elsewhere in Mesoamerica. This very concept underlies the New Fire ceremony of the Aztecs, for example, when every fifty-two years (that is, a full Calendar Round) the world was reborn anew. Maya period endings occurred with far more frequency, but I see similar principles at work in their symbolism. As we have seen, many ritual performances were explicitly cited as re-creations of events of the far distant past, when remote gods dedicated stones or oversaw their own period endings, sometimes billions or trillions of years ago. Calendar rites tied to the Long Count were always repetitive in nature, forever folding past and present, almost as if they were constant efforts to bring Creation into the world of the contemporary.

Like the Maya ritualists of today at a Ch'a Chaak ceremony, ancient kings engaged with the sacred in order to make gods and ancestors present, in order that they could engage directly with people. For this reason, much of their royal ritual involved important metaphors of supernatural birth. We see this perhaps most commonly in the iconography of snakes and serpents. These supernatural "great snakes," as they were called, served as conduits for communication with the supernatural. Their open maws often hold images of people or gods who are "born" or conjured by a king or some other noble person, often as part of a period-ending rite. A stela from Copan shows a standing king cradling a two-headed snake in his arms; from the open mouth of the snake come the two gods known as the paddlers, the two beings who played such an instrumental role in the dedications of stones at the time of the 3114 BC Creation

The "fish-in-hand" glyph, read tz'ak, *"to con-jure."* (Drawing by the author)

event. I suspect that this sort of image, so typical in the art of Copan and other sites, symbolizes the king as the wielder of the sacred, bringing the gods into the present world through bloodletting and prayer.

Texts that accompany such odd scenes of giant two-headed snakes refer to the act of "conjuring" gods (*k'uh*) and ancestral spirits (*k'awiil*). The hieroglyph used to write this action is the "fish-in-hand" sign, which, before it was actually deciphered, was long known to be a glyph associated with bloodletting and other ceremonies of sacrifice. Not long ago, I and others finally deciphered this curious glyph as the word *tzak*, meaning "to conjure something from nothing." As I later noticed, the same word, *tzak*, exists in modern Tzeltal Mayan, used for counting the number of times one grabs a fish out of the water with one's hand, thus explaining the form of the sign. Evidently, "conjuring" a god was likened to the task of wrenching an elusive, slippery fish out of the water, from one realm into another.

Fundamentally, then, the ceremonies associated with period endings were reenactments or re-creations of ritual performances from the far, far distant past, when ancestral gods brought forth spirits and deities into their own "real" world. The very process of dedicating and consecrating stones, stelae, and the like similarly reproduced the very actions of the gods at the moment of Creation at 13.0.0.0.0 4 Ahaw 8 Kumk'u.

I'm reminded now of a statement by the anthropologist Rafael Girard, who wrote some of the most valuable and insightful analyses of recent Ch'ortí Maya ritual he witnessed in the 1950s, much of it now lost. In discussing the significance of their New Year rites, he describes almost exactly what I perceive in the Classic Maya notions of renewal and rebirth at the time of period endings: "On this first day of the year, commemorative of the primordial act of creation, the gods are reborn and act anew, as do their terrestrial representatives and the calendric elements they personify."[3] I couldn't come up with a better characterization of an ancient Maya calendar ceremony, where the ancient *k'uhul ajaw* figuratively gave birth to gods, ancestors, and manifestations of the sacred. He (or she) was also the embodiment and personification of the calendrical cycles being celebrated.

◇

Anthropologists and other academics in the humanities nowadays love to talk about "agency," basically referring to the ability or capacity of a person to act and do things. Maya kings seem to have had lots of it, but there were, I think, some important limits to the idea that their agency was boundless and pervasive, at least when it came to the calendar. It has been said, for example, that rulers in some way controlled or manipulated time and its passage. To some extent this is accurate, for we have clear evidence that K'inich Janab Pakal and other kings "oversaw" period endings. Yet we have to be very precise about how we characterize this relationship. To simply claim that kings "controlled" time reflects a misguided or at least incomplete line of thinking about the all-important intersection of calendar, ritual, and politics.

The notion that any one ruler would be able to exert control over time presupposes that he or she existed outside of time in some way, pulling strings or otherwise acting upon the cycles of time and their operation. I would argue it differently and suggest that the Maya were well aware that such oversight and control weren't possible, that no being, not even some semi-divine actor, could manipulate or disrupt the massively interlaced structure of the calendar. They could "become" time through visual metaphor and through their performances, but controlling time and acting over time it is something that never really comes up in the rhetoric of their texts.

The possible exception to this, to play my own devil's advocate, are instances where kings are said actively to "complete" time periods, if our translations are in fact accurate. In the ancient texts the operative word is *tzutz*, a verb usually translated as "to end" or "to finish," and represented by a glyph showing a pointing hand. Following years and years of scholarly convention, I've routinely interpreted it this way, as in "he ends the ten k'atuns" or "it is the completion of thirteen bak'tuns." (No doubt such wording still plays a significant role in fostering the idea that the world itself will "end" with the coming of the bak'tun in 2012.) There might be other meanings at work here, however, and exploring the subtleties of that language will help us to see just what the ancient Maya thought about units of time and its transitions.

As far back as the 1920s the *tzutz* hieroglyph was recognized as signaling the "end" of a period, and for this reason it widely came to be known as the "completion hand." In the 1980s, I was happy to finally be able to decipher this glyph as *tzutz*, based on phonetic clues offered by the alternative spellings (for example, *tzu-tzu* as a direct replacement for the hand). *Tzutz* is a very appropriate reading, because it means "to stop up, close" in Yukatek, and to "to finish, end, close" in languages more closely related to the glyphs. The sense here is similar to the English expression of "wrapping up" something. But there is more to the meaning of *tzutz* that I find revealing, and that may point to a more nuanced understanding of the term in the context of time. In other living Mayan languages closely related to the ancient texts, *tzutz* also carries the specific meaning of "to redo" some task, mostly in the sense of "to replant." For example, in Ch'ol Mayan, one might say:

Yom lac cha' tʃutʃ jlni cholel ba' ma'anlc tʃa' paʃi ixim.
We are re-planting the cornfield where the maize didn't sprout.[4]

In the Ch'ortí language, the sense of *tzutz* is again "to repeat, do over, replant." It refers to some action that requires more effort in order to be finished, conveying the sense that such things are expected to be done again and again, as in the planting and tending of a cornfield.

I suspect that this takes us a bit closer to the true philosophy underlying the ceremonies of period endings. Kings did not simply "end" periods of time such as k'atuns and tuns; they "replanted" or "repeated" them, in the sense that they actively tended to the periods to ensure their proper coming and going. The word points to the idea that the

The Tablet of the 96 Glyphs from Palenque, recounting the local dynastic history of the seventh and eighth centuries AD. (Photograph by J. Pérez de Lara)

passing of a k'atun is one stage in a sequence of many such passings in the past and the future. When a Maya king "completed" a period of the calendar, he was participating in a long chain of similar kinds of transitions, stretching almost as far as one could imagine. The idea echoes one of the basic concepts of Mesoamerican history and cosmology, in fact: that time and human action are but part of a larger cyclical structure with inherent repetitions. Divine kings such as Pakal the Great did not "end" time in their rituals, therefore; we can say, rather, using a basic agricultural metaphor, that they perpetuated it through "replanting."

To illustrate the richness of the poetics involved, let's look at the opening passage of the beautiful Tablet of the 96 Glyphs from Palenque, a historical text that relates ceremonial events in the lives of several rulers. The tablet was found within the site's main palace, adjacent to the throne room built by the great king K'inich Janab Pakal. His great-grandson commissioned the tablet to commemorate Pakal's one-k'atun anniversary as king; the event was celebrated near the sacred spot where another Pakal also sat as Palenque's greatest king a century earlier. The beginning passage of the text serves as an anchor for the entire text, telling us that the earlier king oversaw the completion of 11 k'atuns, on the period ending 9.11.0.0.0 12 Ahaw 8 Keh.

Lajka Ajaw Waxak Chak Sihoom
Tzutzuy u buluk winikhaab(?)
U kahjiiy K'inich Janab Pakal, aj ho' ? naah, K'uhul Baakal Ajaw

On 12 Ahaw the Eighth of Chaksihoom
The eleventh k'atun ends;
He tends to it, K'inich Janab Pakal, He of the Five Palace
    Houses, The Holy Lord of Palenque.

Here, as in many other texts, the king is said to "tend" to the period, much in the way a farmer tends to his plantings. The word in Mayan is *chabi*, "to do a cornfield," which is in turn based on the word for "earth." The metaphorical relationship points to a certain amount of control over the k'atun, but not so much an overt manipulation of time. It's more accurate to say that descriptions of ancient calendar ceremonies such as this convey the idea that a king watched over time's eternal passage, ensuring that it was properly acknowledged and cared for. In a sense, the king's "cornfield" was time in a cosmic setting.

This is not to say that kings were high-ranking calendar priests. Far from it. In fact, the historical and ancient sources make it very clear that a significant barrier probably existed between ruling officials and those we would call daykeepers or calendar priests. No *k'uhul ajaw* from ancient times ever bore either occupational term. Such priests were no doubt important members of royal court society, or in the larger community as a whole, but they did not rule; nor did rulers assume their specific esoteric responsibilities. I think it's far more accurate to view kings as existing as an integral part of time, not just as outside shepherds or manipulators of days and years. As an *ajaw*, the king was the bodily manifestation of the period, so his responsibilities were in a real sense directed toward an extension of his own self and official identity. One of the king's chief roles was in the performance of time, embodying and symbolizing calendar periods to make them ritually relevant and somehow real for the community at large. The ritual engagement with time was in some ways a social event, an interaction between the king and the animate beings who, besides him, embodied the periods and eras.

Stela 31 from Tikal. (Photograph by the author)

A remarkable illustration of the idea of animated time comes from an ancient text from Tikal, the great capital located near the center of the Maya region, in what is now northern Guatemala. In 1959, excavators there discovered a beautifully preserved stela in a buried shrine, carved on all four sides with an image of a local ruler and his ancestors. The back of the monument bears a long and elaborate inscription recounting the deeds of the kings and many other forebears, focusing in particular on their period-ending ceremonies over the previous centuries. This stone, known today as Stela 31 from Tikal, was itself erected on one such calendar station in the year AD 445, during the reign of the ruler named Siyaj Chan K'awiil (Heaven-born K'awiil). In the Maya system, this was 9.0.10.0.0 7 Ahaw 3 Yax.

This is what we would call a "half-period" station, because the ten in the tun's place indicates that it is exactly one half of a twenty-year k'atun. The Maya used the word *tahn lam,* "center-diminishing" or "half-diminishing," to describe this, a term roughly akin to what we mean when we say something is "half over." But the amazing thing here is what the scribe writes next. Instead of simply saying that the current k'atun time period is half over, the writer goes into remarkable detail, listing the names of several important gods and deities, many of whom remain obscure. Heading the list are "the eight thousand heavenly gods, the eight thousand earthly gods," collective terms for the multitude of deities in the ancient Maya pantheon. Various names follow, including the familiar paddlers, who are known to "oversee" and witness period-ending rites by rulers all over the Maya area. But there are a number of others, including a revealing term *Bolon Tz'akb(il) Ajaw,* which refers collectively to the divine royal ancestors of the dynasty (literally it means "the Many-ordered Lords").

So, Stela 31 is explicit in telling us that the gods and ancestors listed *are* time. It is the totality of cosmological beings that are, in some way, "half over" when the midpoint of the k'atun era is reached. What does this mean? If I might speculate a bit, I would hazard to guess that each ka'tun was thought to encompass and manifest a microcosm of the sacred. Each period was a world unto itself, with its gods and forces of nature ever renewed and reborn through the efforts of kings and their proper ceremonies. Let's recall that k'atuns were the most important units of time when we think of long-term human experience (the next highest period, the baktun, was far too long, at about four hundred years), so we might well understand how each twenty years, the gods and time itself were given a new lease on life.

The poetic language of Stela 31 goes further in directly equating the existential quality of the gods and the ancestors with abstract notions of time itself. When the k'atun is half done, so are the sacred beings who define the very world. And further still in this text, we read how Siyaj Chan K'awiil is following in the footsteps of his predecessors, continuing a long line of work on the various k'atun stations. The imagery evoked is of laboring in an everyday cornfield, sewing and sustaining the k'atun periods as if they were sacred maize plants.

Prophecy as History

It's no exaggeration to say that Maya archaeology and research began at the ruins at Copan, located in a small green valley in western Honduras. It was there that John Lloyd Stephens first encountered a major ruined city during his famous travels in the region in 1839. His bestselling book, with its wonderful engravings of Copan's ornately carved stelae by artist Frederick Catherwood, captured the imagination of generations of future travelers and archaeologists. Numerous archaeological projects followed and have continued up to the present day, making Copan perhaps the most intensively studied and excavated of Maya ruins. The ancient city is defined mostly by a large acropolis, a concentrated mass of pyramids and courtyards built on the flat plain of a small valley. The well-preserved stelae made famous by Stephens and Catherwood stand like statues in the large open plaza to the north, all inscribed with period-ending dates and descriptions of the rituals undertaken by the kings on those days. The pyramids to the south are largely crumbled, but their slopes are covered with the sculpted stones that decorated their façades, mostly with cosmological symbols evoking the sky, sacred mountains, and more complex symbols relating to sacrifice and royal ancestry. The impressive architectural remains of the city indicate that

Copan's Altar Q, showing sixteen rulers oriented to the four cardinal directions.
(Photograph by the author)

it was an artificial sacred landscape, much as we see at Palenque and other major Maya centers. At Copan, however, the focus of the religious and dynastic symbolism was mostly on one man, K'inich Yax K'uk' Mo' (Great Sun Green Quetzal-Macaw), an ancestral founding father who was at the center of a dynastic cult for centuries of Copan's history.[*]

One stone sculpture from Copan neatly summarizes this ancestor's importance. Altar Q is perhaps the most important single stone monument from the site, although it is not terribly imposing or physically impressive. The box-shaped stone originally stood on four legs in front of the temple shrine dedicated in AD 775 as a monument to K'inich Yax K'uk' Mo', who had lived nearly 350 years earlier.[†] Sixteen seated fig-

---

[*] See Stephens, 1839 (in various modern reprints). Two excellent and accessible introductions to the archaeology of Copan are by Fash, 2001, and Webster, 1999.

[†] The founder's likely tomb was discovered by archaeologists Robert Sharer and David Sedat directly beneath this temple, deep down among the early phases of the acropolis dating to the early Classic period, in the fifth century AD. See G. E. Stuart, 1997.

ures, all dressed as rulers, decorate the four sides of the square stone, four people to a side. On the front, or western, side we see two of the lords facing each other, with two small hieroglyphs in between them recording the Calendar Round date 6 Kaban 10 Mol. This is a famous date in Copan's dynastic history, corresponding to the crowning of the last firmly known king of Copan, Yax Pasaj Chan Yopaat. (The full date is 9.16.12.5.17 6 Kaban 10 Mol, or June 28, AD 763, somewhat close to summer solstice.) All of the royal figures on Altar Q sit upon large hieroglyphs that give a personal name or title, allowing for their easy identification as the sixteen kings of the local dynasty. The sequence begins with the figure at left of center on the altar's front, a portrait of the dynastic founder himself, K'inich Yax K'uk' Mo'. He sits upon another glyph that simply states "*ajaw*," "king." The sequence runs counterclockwise around the stone, ending with the sixteenth ruler, Yax Pasaj Chan Yopaat, who faces his remote predecessor, receiving from him a stafflike instrument of office. The date glyphs between them leave little doubt that this is a metaphorical scene of the contemporary king's accession, as he receives sanction and blessing from the long-dead founder, buried deep below the site of the temple and its altar.

I know of no other monument like Altar Q, with its visual presentation of collapsed history. Its message of a royal gathering or meeting spans nearly four centuries of kingship.* The current ruler assumes his office directly from the founder as all of his other illustrious predecessors look on in approval. Even at Copan, with its hundreds of major monuments dedicated to kings and their ceremonies, Altar Q stands out as a cohesive and powerful statement about the power of royal ancestry and its key role as an underpinning of rulership and its ideology.

The altar is also, I think, much more than just a visual king list. Its design emphasizes continuity and communication over the centuries, yet also molds history into a firm, four-sided cosmological framework.

---

* Before the historical interpretation of Altar Q, it was for many years seen as representing a meeting of astronomers, based on a rather fanciful interpretation first put forward by Teeple (1930). Once again we see how powerful a hold astronomy and calendrics had over Maya scholarship of that time.

The absolute symmetry in which sixteen rulers are shown equally divided into four world directions is inescapable and certainly intentional. This, on the face of it, may not seem so surprising—after all, Maya kings and dynasties were always keen to place themselves in a cosmological setting—but what I find remarkable is the way this four-by-four closed system reflects the actual historical development of Copan's dynasty. Yax Pasaj Chan Yopaat was Copan's sixteenth and last ruler, and many of his inscriptions feature this fact as an honorific title, "the Sixteenth Successor." After him we have no firm record of any king. What's more, some buildings of his royal compound were burnt before they collapsed in a heap of stone, suggesting a deliberate destruction to the king's residential area. It would appear that soon after the reign of the sixteenth king, or even during it, Copan's dynastic center was abandoned and left to decay, probably sometime in the early ninth century. All of this makes me wonder if Altar Q's design and message represent the Maya idea of a historical totality, perhaps even showing us a very conscious and deliberate idea of a dynasty that has come full circle, with a beginning and an end.*

Once again, time may have much to do with this sense of historical closure. According to the history recorded on the top of Altar Q, the founder K'inich Yax K'uk' Mo' arrived in Copan in AD 427, after a long journey from a distant ceremonial locale, perhaps Teotihuacan. Wherever his starting point, he "received k'awiil" at that highland Mexican city, evidently an initiation ritual of some type. His arrival to Copan came on 8.19.11.0.13 5 Ben 11 Muwan, only several years prior to a Long Count station of great significance: 9.0.0.0.0, coming in the year 435. According to many later Copan texts, K'inich Yax K'uk' Mo' oversaw this key period ending along with his son, whom we simply call Ruler 2. A monument showing the father and the son was dedicated in the plaza below the acropolis, where, generations later, a tall

---

* One possible reference to a later king named U Kit Took' appears on Altar L at Copan, a very late and unfinished monument on which he is portrayed together with Yax Pasaj Chan Yopaat. There are a number of ambiguities with Altar L, and I'm not convinced as yet that U Kit Took' was ever a king. Perhaps the discovery of another late inscription at Copan will someday clear matters up.

stela was erected to commemorate the same date, in retrospective fashion. All signs point to the bak'tun ending being the defining moment in the career of K'inich Yax K'uk' Mo' as the founder of the dynasty.

His distant successor in office, Yax Pasaj Chan Yopaat, ruled from 763 to at least the year 810, when we find the very last mention of him in the historical record. This came only twenty years before the end of the next bak'tun, and a mere eleven years before the one-bak'tun anniversary of the founder's arrival in Copan. In other words, 95 percent of the entire span of the bak'tun from 9.0.0.0.0 to 10.0.0.0.0 neatly encompasses Copan's ruling dynasty; the final 5 percent represents a lack of any evidence whatsoever. This correspondence between the beginning and ending of the bak'tun with the rise and fall of Copan's rulers is striking, and very unlikely to be coincidental.

I suspect that the arrival of K'inich Yax K'uk' Mo' in Copan was intended to establish a new ritual center predicated on the soon-to-occur turn of the bak'tun on 9.0.0.0.0, in AD 435. As a small community, Copan had existed long before this time—its archaeological remains go very deep in parts of the valley, in fact—but before the fifth century, it never participated with the high rollers of Maya civilization. In the Late Classic era it was a great center of elite power, but, in general, the city of Copan was a latecomer on the scene when compared with other major cities such as Tikal and Calakmul. Significantly, different lines of evidence also show that K'inich Yax K'uk' Mo' was not originally from Copan, and that he likely spent most of his childhood in the central Maya lowlands, perhaps in the general vicinity of Tikal or maybe in what is now Belize. Unfortunately we know nothing about what political and religious decisions might have gone into his arrival at Copan. Who called the shots? To whom did K'inich Yax K'uk' Mo' answer in the hierarchies of Maya politics at that time? We can't know. All we can do is read between the lines, and take note of the fact that he was forever linked in Copan's history to the great 9.0.0.0.0 date—a period ending that was for all intents and purposes Copan's opening act.

At the opposite end of the site's history, we see a large and densely populated community beset by political and demographic problems. By AD 800, some twenty-five thousand souls were living in and around the Copan Valley, burning precious wood and occupying the valley's valuable arable land. Disease was commonplace, and the power of the kings was evidently waning as more junior members of Copan's royal court gained political power and prestige. As was true all over the Maya region and elsewhere in Mesoamerica at the time, population pressures were making life fairly miserable for the commoners of the area, and they were probably taking their toll on the ruling class as well. While all of this was happening, it could not have been lost on members of the Copan royal court that the next bak'tun ending was fast approaching. Not only that, but Yax Pasaj Chan Yopaat was the sixteenth in a series of kings who, according to Altar Q, had established roles in the cosmic order. To me, the symmetry of that stone's depiction offers a sublime statement of closure, of a certain beginning and a clear end point. I have to wonder, then, how could Ruler 17 of Copan, if he ever existed, have fared at all? Was the dynasty of Copan a foregone conclusion, based on the structure of the calendar and the numbers of kings? Such a scenario seems overly simplistic, to be sure, not to mention deterministic. But there seems to be something to it, I keep thinking.

Copan is unique in structuring its royal history within such well-defined brackets, each tied to a key transition of the bak'tun. But other sites appear also to have patterned their dynastic histories in ways that suggest some amount of foreknowledge that beginnings and endings were predetermined. At the other end of the Maya world, at Palenque, for example, the sequence of kings conforms to a strange and fascinating pattern. Again we see sixteen rulers in the list—perhaps a coincidence. The founder of the dynasty was named K'uk' Bahlam (Quetzal Jaguar), and like his counterpart in Copan, he is said to have ruled at the time of the 9.0.0.0.0 bak'tun ending in AD 435—again, perhaps, a coincidence. K'uk' Bahlam may have had his early royal court at a location different from that of the Palenque we know today, which was later established in the year 490 as the seat of the dynasty, under the reign of another early king, Butz'aj Sak Chihk. The younger brother of this king, in turn, later became an important ruler in his own right, and was likely to have been a direct ancestor of the famous

The Rulers of Palenque

RULER NAME

K'uk' Bahlam

Ch'a ? II

Butz'aj Sak Chihk

Ahkal Mo' Nahb

K'an Joy Chitam

Yit K'uhil

Kan Bahlam

Ajen Yohl Mat

Janab Pakal

Ix Yohl Ik'nal

? Muwaan Mat

K'inich Janab Pakal

K'inich Kan Bahlam

K'inich K'an Joy Chitam

K'inich Ahkal Mo' Nahb

Upakal K'inich Janab Pakal

K'inich K'uk' Bahlam

Palenque ruler K'inich Janab Pakal, who would come onto the scene a century later. The entire dynasty founded by Kuk' Bahlam lasted into the eighth century at least, into the reign of the very last Palenque king we know, named K'inich K'uk' Bahlam, for his great ancestral founder. As we look down the full name list, we see that some other names also repeat. And there is an odd, wonderful pattern to the names. Five of the early dynastic names repeat themselves in the Late Classic, but in *precise reverse order*. The first king to reuse a predecessor's name this way was K'inich Janab Pakal himself, who took the name of a man who may have been his grandfather. Pakal's two sons (each reigned as king) also took the names of earlier lords. Finally, the last known king followed suit and assumed the name of the celebrated local founder.

Here we see another sense of closure and finality, different from Copan's in many ways, but nonetheless looking very self-deterministic in its structure. Palenque's later kings, it seems, deliberately chose to "fold" time back on itself, and repeat the sequence of the kings who came before them. With some other names intervening (for history is seldom so cleanly structured), we have the pattern: 1-2-3-4-5, 5-4-3-2-1. Again I have to scratch my head and ask: Were the Maya consciously portraying their dynasties as "closed systems" of some sort, with beginnings and endings, and following an internal symmetry?

The notion that Maya history was to some extent determined by the calendar's prophetic structure, and not just measured by it, isn't at all new. One of the great Mayanist historians of the twentieth century, Ralph Roys, raised a similar point based on his profound knowledge of the Books of Chilam Balam and their depiction of native history according to the cycles of the k'atuns. As he wrote many years ago, "The events recounted in the Maya Chronicles . . . offer excellent grounds for believing that this belief was so strong at times as to actually influence the course of history."[5] And my late friend Dennis Puleston proposed an even more radical notion based on the same idea, suggesting that the famous Maya collapse itself could be explained in part by the built-in, self-fulfilling nature of Maya cyclical history. To Puleston, the collapse may well have been "fully anticipated by ancient Maya scholars and priests, who by means of consultations with their books and prophecies were well aware of their impending fall." Both Roys and Puleston had no access to the deciphered hieroglyphic texts of the Classic period; Puleston's article on prophetic history was published in 1979, just on the cusp of the first great advances in glyph decipherment.[6] Yet their keen sense of the predestination of events certainly rings true with what we see with the curious dynastic histories at Copan and Palenque.

Interestingly, few archaeologists hold much stock in the idea that the multiple rises and falls of the ancient Maya might be closely tied to these inner workings of the abstract calendar and concepts of prophecy. The collapse of Classic Maya culture is widely seen instead as a systemic failure of rulership and authority, where factors such as population pressures, warfare, and environmental degradation brewed together and reached a critical mass, forcing thousands of people to flee cities and towns, leaving the ideology of kingship—and the Long Count calendar so closely tied to it—with no reason to continue. I agree with this scenario in many ways, but I'm also inclined to think that the mechanisms of time were also part of this fateful mix. The Classic Maya must have been very conscious of the way the approaching end of the bak'tun seemed to coincide with profound societal pressures and logistical difficulties, and perhaps to some the coming of a new era was a way to understand and account for them. Once time has explanatory power, it can take on a life of its own, and self-fulfilling prophecies can't be too far behind. No matter how we eventually come to account for the end of Classic Maya civilization in the ninth century, it seems that the vast Long Count calendar played a role in politics and religion. To me, it probably is no accident that after the collapse of scores of kingdoms in the ninth century AD, the Long Count essentially ceased to be a prominent part of ancient Maya cosmology.

There's another point to be made here as we approach our next and final chapter. Many people today see the ancient Long Count calendar as having a profound relevance for our own modern world, as a mechanism that predicted an upcoming "end of times" or a "transformation of consciousness." It's as if the ancient Maya somehow could anticipate the fears and struggles we experience in our modern industrial life, and offered a mystical end game we could look forward to. No such luck. The truth of the matter is that the Maya calendar was inseparable from the ancient world that created it: a lost worldview of kings, gods, and ancestors. By wrenching this special vision of time and cosmology away from that particular cultural and historical milieu, we do nothing more than manipulate the past for our own purposes and messages. Modern society always does this sort of thing with the symbols of past civilizations, but, as our final discussion hopefully demonstrates, the ancient Maya and their modern descendants deserve better.

10

SEEING STARS

If we regard millennial passion in particular, and calendrical fascination in general, as driven by the pleasure of ordering and the joy of understanding, then this strange little subject—so often regarded as the province of drones and eccentrics, but certainly not of grand and expansive thinkers—becomes a wonderful microcosm for everything that makes human beings so distinctive, so potentially noble, and often so actually funny.

—*Stephen Jay Gould* [1]

For more than a century the Maya have held a singular place in the popular imagination as one of the weirdest, least accessible, and most mysterious of ancient civilizations. Open up an old guidebook or text from the 1930s or '40s touching on the Maya and you are bound to find references to the "mystery" of Maya ruins and the unusual intellectual culture of their builders. The Maya were often portrayed back then as a people oddly obsessed with time and its passage, always peaceful and aloof, and forever gazing skyward at the stars and planets. Needless to say, this view changed dramatically in later decades, in the wake of sophisticated archaeological research and the decipherment of ancient texts. The sudden revelations about history and kingship were especially important in showing how real-world themes of politics, ritual, and war were central in Maya history and culture. In this way, the popular image of the Maya and other Mesoamerican cultures now emphasizes bloody warfare and ritual at the expense of the "peaceful" model of decades past. Nowadays, opening up a magazine or going to the movies, we're more likely to see the

Maya portrayed as bloodthirsty priests and warriors, as best exemplified in Mel Gibson's strange and jarringly violent epic film *Apocalypto*. It makes me wonder if the pendulum has perhaps swung too far in the other direction.

It's not so easy to dismiss the naïveté of those early scholars who misunderstood the reality of ancient Maya culture and its history, and who preferred a more romanticized vision of an advanced, intellectual, and peaceful civilization in the rain forest. In those days of world wars and the Great Depression, such views might even seem desirable and understandable. Today's opposite representation of Maya culture as strangely obsessed with violence, in a way far beyond other ancient peoples, is, I suggest, equally an artificial creation, rife with assumptions and inaccuracies. And whereas both views might seem contradictory at first, I sense that they each reflect a conundrum long faced by Westerners, and North Americans especially: to understand the Maya as true human beings. For many in our modern society, the ancient Maya still represent a supreme example of a mystical, otherworldly people, perceiving cosmic truths that we ourselves can't, arriving on the scene and then "disappearing" suddenly in a mysterious fashion. Their calendar and prophecies remain strange and alluring, almost alien in nature.

I'm often reminded of this during conversations at cocktail parties or in airport lounges, once people learn that I'm an archaeologist. The queries quickly come flowing out: "So what happened to the Maya anyway?" "Why did they all disappear?" or "How could they have built pyramids without metal tools, or wheels? Does it mean they visited Egypt?" or "What was with all the human sacrifice?" Tackling these sorts of innocent questions is usually an engaging and enjoyable exercise, but it isn't always easy, given that such deceptively simplistic questions are often based, at least in part, on valid academic debate and discussion. The Maya didn't "disappear," of course, since five million speakers of Mayan languages are still with us. (I mean this quite literally: many Maya now live in California, Illinois, North Carolina, New York, and throughout the United States.) Sure, a great many of their ancient cities were abandoned in sudden fashion around AD 800, and we archaeologists have been debating why for decades. Before and after this "collapse," the Maya made many pyramids roughly comparable to those in Egypt, not because of any long-distance cultural contact between Mesoamerica and the Near East, but because such pointed architectural forms, wide at the base and narrow at the top, were the only means of making a tall building before the advent of steel skyscrapers. (Their having only "stone age" technology didn't matter, since one can make perfectly serviceable and even superior tools out of flint or obsidian.) Pyramids appear the world over at different times and places, from the Americas to Asia, and often with surprisingly similar functions, as artificial mountains and coverings for elite burials. But what accounts for this near-universal tendency in the ancient human imagination to create artificial landscapes? Simple questions therefore can lead to awkward answers that may or may not prove very satisfying. At times I can't help but feel that such answers might disappoint, since the reality strips a good deal of the allure out of our highly romanticized visions of the Maya as somehow different from other ancient civilizations. In a way, and for very complicated reasons, we do like the "mystery" of the Maya, and probably will continue to for some time.

Origins of a Mystery

The perception of the Maya as somehow unique or out of place among other ancient cultures has a long intellectual pedigree, ultimately traceable, I think, to the earliest speculations by European settlers concerning the origins of Native American peoples. Prominent intellects of early America, Spanish- and English-speaking alike, were adamant in their belief that the natives of the New World—the mislabeled "Indians"—were descended from the lost tribes of Israel. Both Catholic Spaniards in Mexico and Puritans in early North America relied on the Old Testament as a basic historical tract, after all. One early writer who held such views was the priest Fray Diego Durán, who lived near Mexico City in the sixteenth century and who left us, among other important documents, his important treatise on Aztec festivals and timekeeping, *The Ancient Calendar*. Durán's deep knowledge of Aztec history and culture led him to ponder many apparent similarities between the Aztec ways and Christian and Jewish traditions. Similarly, in early America, writers such as Thomas Jefferson considered a Jewish origin for the American Indian population; Jefferson even instructed his friend Meriwether Lewis to keep a keen eye out for firm evidence of such a connection during his travels with William Clark in the unexplored American West.*

Given the intellectual currents of the time, it's easy to understand how many early explorers and writers in the 1820s and '30s could so summarily reject any historical connection of the great ruins at Palenque or Uxmal with the Maya people who lived nearby; clearly, the thinking went, these impressive buildings must have been built by others more skilled and civilized who had arrived in America in the remote past from the Old World. Even Joseph Smith's writings and the Book of Mormon reflected the widespread notion that Native American peoples and Mesoamerican ruins arose from outside, very distant origins. According to the tenets of the Mormon faith to this day, most Native American populations are believed descended from colonists known as "Lamanites," who centuries ago overran the more civilized "Nephites," who built the great stone cities seen in Mexico and Central America.†

One early advocate of this general view was the Englishman Edward King, Lord Kingsborough, who was the first to publish the ancient Maya book known as the Dresden Codex in his monumental work *Antiquities of Mexico*, printed in nine volumes between 1830 and

---

* For Durán's diffusionist views, see commentary and edited works in Horcasitas and Heyden, *Book of Gods and Rites and the Ancient Calendar*, pp. 23–31. Jefferson's interest in the Lost Tribes of Israel in connection with the Lewis and Clark expedition is mentioned in Ambrose, 1996, p. 154.

† An overview of views of early scholars and explorers who saw the distant origins of the ancient Maya and other Native American peoples are discussed by Wauchope, 1962, and more recently in Evans, 2004.

1848 (each massive volume of this magnum opus weighs nearly thirty pounds!). Like many of his contemporaries, Kingsborough was convinced that the ancient builders of the "Mexican" monuments must have been Israelites. Another eccentric European of the nineteenth century, an explorer and artist named Jean-Frédéric Maximilien de Waldeck, held vaguely similar diffusionist ideas about the origin of the ancient Maya, although without direct reference to the Old Testament. His published illustrations of ruins in Yucatán and Chiapas from the 1830s through the 1860s include many details that evoked Egyptian and Asian styles and motifs, and went along with his claims that the builders of Palenque, for example, belonged to "the white race."

And so we see how early American intellectual culture and its perception of history set the stage for later thinking on the subject, and probably influenced some popular perceptions held today. Throughout the nineteenth century, the Maya seemed incompatible with long-standing and even ingrained preconceptions about Native Americans in the expanding United States. In popular culture and the media we continue to wrestle with understanding Mesoamerican civilization and its remote origins. All is mysterious still: The Maya left lofty pyramids in the remote, inhospitable jungle, and abandoned them so suddenly and completely. They wrote with strange-looking symbols that few could read. Their art emphasized odd, baroque-looking designs coupling images of fierce kings with serpents and grotesque gods. These ideas have contributed to the romance of Maya and Mesoamerican archaeology since before the days of Stephens and Catherwood in the 1840s.

Today attributions of a certain uniqueness or exotic quality to the ancient Maya and their intellectual culture, especially where astronomy and the calendar are concerned, are still everywhere to be seen, whether in the scholarly literature or in the writings of the New Age fringe. It isn't hard to see why. When in the early twentieth century it seemed that the decipherment of hieroglyphs was a distant or unreachable goal, many saw the calendar and astronomy as the only available inroads into understanding the intricacies of Maya script. As we've seen, Sylvanus Morley, the archaeologist and one-time OSS agent who dominated Maya studies between the wars, was obsessed with finding

An early representation of a Palenque sculpture by Jean-Frédéric Maximilien,
Frédéric Comte de Waldeck, emphasizing the Greco-Egyptian style. Like many
intellectuals of the early nineteenth century, Waldeck was reluctant to believe that
ruins of Mexico and Central America were built by Native Americans.

new inscribed monuments, sometimes in very remote places. But he had an interest in finding only Long Count dates he could read, and therefore gave little regard to the (to him) unreadable strings of glyphs that, as we now know, provided the meaningful historical core of the texts themselves. To Morley, like many of his day, reading bars and dots was the essence of epigraphic research, in order to discern dates, moon ages, and maybe other astronomical lore. Before the 1960s, when the existence of history was finally established in Maya inscriptions, epigraphers were confronted by long strings of dates that Morley, Thompson, and others had decoded. What were the dates doing, if not recounting history? Writing in 1922, Morley was adamant that astronomy was the be all and end all of Maya inscriptions:

> No grandiloquent record of earthly glory these. No bombastic chronicles of kingly pomp and pageantry, like most of the Assyrian, Babylonian, and Egyptian inscriptions. On the contrary, the Maya priests would seem to have been concerned with more substantial matters, such as the observation and record of astronomical phenomena.[2]

Even now, the ancient Maya are regularly singled out and celebrated for their astronomical aptitude. My colleague Anthony Aveni, a noted scholar of Maya astronomy, has said, "Nowhere were the qualities of advanced civilization and intellectual achievement more outstanding than in the land of the Maya."[3]

The Maya were fine astronomers, of course, but I find it interesting that, apart from routine notations of moon ages, the written record of the ancient Maya in fact contains very little astronomical record-keeping. This is a something of a radical statement on my part, given the prominence that astronomy has been given in Maya hieroglyphic research over the past one hundred years, even until recent times. But everything changed in the wake of Proskouriakoff's great breakthrough of 1960, and with the advances we've made in decipherment. The ancient inscriptions actually *do* concern the "grandiloquent" matters of kingship and royal ritual, and pay virtually no attention to astronomical phenomena beyond recording some dates in the lunar calendar.

We do see astronomical records in the precious Maya books that have survived, such as the Dresden Codex, with its tables and charts revealing the Mayas' careful observations of the major planets such as Venus, or of eclipse phenomena. And the Maya recognized a number of significant stellar constellations—not necessarily equal to our own—set against the dark firmament of the night. To the Maya, the starry sky was a nocturnal landscape full of animate movement and divine meaning. But in this they were no different from other Mesoamerican peoples who possessed similar knowledge and awareness of the mechanisms of stars and planets.

It all makes me think that the common perception of the ancient Maya as uniquely great astronomers is misguided, or that we've oversold the idea that they were especially talented where stargazing and numerology are concerned. We do well to remember that the Long Count calendar—perhaps the most awesome numerological analysis of time ever devised—was used first among non-Maya peoples in the Olmec region. And were Maya priest-astronomers better observers of the skies than their Aztec counterparts? I doubt it. It strikes me that our common notion of Maya uniqueness actually stems from the chance survival of certain original sources, most notably the Dresden Codex, with its almanacs and planetary tables. The prominent role of that precious document in the early study of Maya writing (Förstemann used it at his desk to decode the Long Count) has to some extent skewed our vision of who the Maya were, emphasizing their super abilities in astronomical study over those of their Mesoamerican neighbors. In other words, if the Dresden Codex had never survived the ravages of the Conquest, not only would the course of Maya research have been utterly different, but so, too, I suspect, would the image of the Maya in the popular imagination. Surviving pictorial documents from central Mexico lack the same detailed *written* accounts of Venus cycles and tables of eclipses, but they do con-

A page from the Venus tables of the Dresden Codex.

tain very sophisticated numerological content, some of it still poorly understood. This skewed sampling of Meosamerican priests' handbooks doesn't mean that Aztec and Mixtec priests lacked astronomical knowledge similar to that of their Maya counterparts. Different sorts of Venus tables and other planetary records doubtless were kept throughout all of Mesoamerica; it so happens that the one great surviving example is Maya. My sense, then, is that the Maya have been given a bit too much credit for having some monopoly over astronomical awareness and acumen. And I say this as a Mayanist who has long been in love with the culture and its legacy.

Let's take a look at one of the most famous sections of the Dresden Codex, and how it presents information on the cycles of Venus as morning and evening star. It's important to stress that the Venus tables of the Dresden Codex are not simply tallies of observed data like we might see in the notebooks of, say, Renaissance astronomers. They are instead approximations of astronomical phenomena originating from long-term real observation, but "tweaked" to conform to other ritual cycles that made the astronomy relevant. In other words, many of the numbers represent a structure recognizable as having a basis in the movement of Venus, but they are at the same time "contrived numbers," meant to show ways that the movement of the planet could in a general way conform to other cycles and important numbers.

The first decoder of the tables was our hero librarian Ernst Förstemann, who by crunching the numbers was able to discern that the writer of the table was trying to emphasize four particular stations of Venus as it rose and fell as morning star and evening star. Along the bottom of the five pages (46–50) he saw the following day intervals, recorded as bar-and-dot tallies in the Long Count system: (a) 236 days, which he related to the time when Venus was visible as a morning star; (b) 90 days, corresponding to the invisibility of Venus at superior conjunction; (c) 250 days, when Venus was again visible, but as an evening star; and (d) a period of 8 days when the planet once more disappeared at an inferior conjunction. Now, these numbers are not exactly an accurate reflection of astronomical reality. The actual observed mean intervals are shown in italics as follows, after the numbers, as they are written in the Dresden Codex:

$$236 \ (263)$$
$$90 \ (50)$$
$$250 \ (263)$$
$$8 \ (8)$$

Quite a discrepancy here, except in the very last number. The mean day intervals are only those derived from observed spans of days that can vary considerably from year to year. As Anthony Aveni has pointed out, these astronomical numbers were modified by the Maya in order to make them tally and intermesh well with other kinds of cycles. The first number, 236, which is related to the visibility of Venus as a morning star, is "off" by 30 days, but it does equal 8 lunar months. Similarly, 90 days conforms to 3 lunar months. For the Maya timekeeper, the point seems to have been to come up with a number that *conceptually* accommodated different types of heavenly phenomena, here using Venus as a frame of reference to represent a certain elegant, even if somewhat forced, symmetry in the skies. Even in this grandest of demonstrations of Maya astronomical ability, we see how the movement of the planets was meaningful only when contextualized in a larger cosmos of gods and numbers.[4]

Early research on hieroglyphs took Förstemann's astronomical observations and ran with them. By 1930, Teeple had worked out the basic elements of a lunar calendar that served, as later research discovered, as a cosmic anchor for records of history and myth.[*] By the 1940s and '50s, Maya culture was naturally described as being obsessed with time, the planets, and the stars. I would argue that this forever gave the

---

[*] See Teeple, 1930, for his work on the Mayan lunar calendar; also useful is Thompson, 1950. A broad treatment of early Mayan glyph research at this time is my father's very thorough article, in G. E. Stuart, 1992.

Maya an air of the exotic and otherworldly in the popular imagination, even to this day.

Star Wars

By the 1970s and '80s the two tracks of research on Maya, one focused on astronomy and the other on history, had begun to converge. In studying the written accounts of wars between cities, for example, it appeared that the timing of important battles may have been directly influenced by the movements of planets and stars. As the many city-states of the ancient Maya political landscape developed and grew over time, warfare became an increasingly common and destructive aspect of politics, economics, and everyday life. Temple art is replete with images of prisoners being taken at spear point, and of tortured captives kneeling before victorious kings. Strangely enough, however, Morley, Thompson, and other early writers were blind to what now seems obvious visual evidence of warfare in Maya culture. In the years in between World War I and II, in particular, they strove to see the ancient Maya as anything but warlike. As Morley himself put it, "True, bound captives are occasionally portrayed, but the groups in which they appear are susceptible of religious, even astronomical interpretation, and warfare as such is almost certainly not indicated."[5] In typical fashion for the time, astronomical readings of Maya art trumped all.

For Morley, who had worked in Yucatán for most of his career, "Old Empire" meant "pre–Chichen Itzá." That great northern center has long stood out among other ruins for its strong central Mexican influences, and for its associated severe style and militaristic theme in its art and iconography. The contrast of the harsh "Toltec" with the elegance of earlier Classic Maya art and culture seemed almost like like night and day to Morley and his contemporaries, and it provided the mental armature, I think, for a "peace then war" mind-set about Classic Maya development over the long haul. It's important to stress how Morley was a tireless advocate for the Maya and Maya research, giving many

The realistic views of warfare and palace life in the Bonampak murals, discovered in 1946, greatly changed scholarly views of ancient Maya culture. (Photograph by the author)

public lectures and writing articles for *National Geographic,* touting the "great intellectual achievements" of his beloved Maya. In an era when an unresolved world war was still looming on the horizon, he simply couldn't see them as warlike.

Morley died in 1948, soon after one of the great discoveries in Maya archaeology would force a major reevaluation of the "peaceful Maya" model. The small ruins of Bonampak, in the jungles of Chiapas, Mexico, had come to light in 1946, revealing a remarkably preserved building with ornate colorful murals. This one structure has left a profound mark on Maya studies ever since, for a cursory look at its narrative content is enough to convince anyone of the violence of Maya war. One of the three painted rooms depicts a massive battle, with spears upraised and bodies flying through the air in combat. Another room at Bonampak shows the judgment of prisoners taken in that same battle, presumably. The terror in the eyes of the tortured captives is palpable. For Morley, the discovery of Bonampak came too late to significantly change his views about the peaceful Maya, but he did see photographs and drawings of the vivid and realistic imagery before he passed away.

Even in the wake of the Bonampak's discovery, it took archaeology some years to recognize that militarism didn't come only with the arrival of foreign warriors from highland Mexico, but that it actually had deep and early roots in Maya civilization. Much of the change came during the 1960s and '70s, some of it related to the quickly changing ideology of American academia in those years of social and political conflict. New generations of archaeologists grew impatient with old-fashioned techniques and ethereal interpretations of culture, and were far more comfortable emphasizing more scientific and positivist approaches in studying the past. The idea of a peaceful Maya world simply didn't ring true anymore, especially after Bonampak and physical evidence of warfare from other quarters. My colleague David Webster documented the existence of elaborate fortifications at the site of Becan and other centers—some very old indeed—giving clear indication that the Maya were far from peace-

A hieroglyph for warfare shows a prominent "star" sign, leading many in the 1980s to propose an astronomical basis for the timing of Maya battles, or "star wars." (Drawing by the author)

ful.* For the field and for the image of the Maya in the popular imagination, this necessitated a jarring adjustment.

As noted earlier, advances in decipherment in the 1970s and '80s added an entirely new dimension to the study of Maya war, giving it historical context. Numerous texts at some sites emphasized warfare over and over again, recording on stone monuments the capture of prominent nobles from neighboring kingdoms. One such glyph for war depicts a star sign above another sign for "earth," flanked by small droplets representing water and rain. By the late 1970s, this glyph, although still undeciphered, came to be called the "Star Wars" glyph. (One can probably guess why, based on the timing.) The star sign was an irresistible clue, suggesting to some researchers that warfare, while real, was more of a ritualized conflict, involving small raiding parties and perhaps timed according to astronomical phenomena. I see echoes here of Morley's old idea that images of war were mostly "religious" and "astronomical" in nature. Some ideas die hard.

Bonampak emerged in the 1980s as a key piece of evidence suggesting that Maya wars were timed to Venus phenomena. Above the scenes of the battle and the judgment of the prisoners are celestial bands that include several "star" figures, almost certainly representing constellations. The celebrated Mayanist Floyd Lounsbury of Yale University studied the hieroglyphic inscriptions of the murals, including the badly preserved large caption text that accompanies the main battle scene. In a celebrated paper, he posited that the date of this battle was written as 13 Chikchan 13 Yax, corresponding to the Long Count date

---

* Webster's emphasis on the role of war in Mayan civilization is best summarized in his excellent book on the collapse, Webster, 2002.

9.18.1.15.5, or August 2, AD 792. Lounsbury was interested to find that this date fell precisely on the inferior conjunction of Venus, and approximated the day of the sun's zenith passage at the latitude of Bonampak.[6]

It turns out, though, that Maya war probably had little directly at all to do with Venus, despite the assumptions that we all made some twenty or thirty years ago. The date glyphs accompanying the battle scene at Bonampak were probably misidentified by Lounsbury, as infrared photographs of the painted glyphs have since revealed. And if one studies the wide sample of "Star Wars" glyphs so far identified in all texts, one finds no discernable connection to the rising or setting of Venus.[*] While the precise reading of the "earth-star" glyph for war remains elusive, its image may be based on the idea of a "raining star"—the flanking dots on either side of earth are clearly images of falling water. In one early example, the star sign sits in a cleft of the earth, suggesting an idea of breaking or opening. The sign suggests the idea of a meteor or falling star, both of which have strong associations with warfare in Mesoamerican thought. Among the modern K'iché' Maya, a meteor is called *ch'olanic ch'umil*, "the star that makes war," and its appearance was seen throughout much of Mesoamerica as an omen of war and destruction.[7] Now, in the last ten years or so, the once-popular notion of a Venus-regulated "Star Wars" among the Maya seems less and less likely.

Maya Cosmos

The study of ancient Maya astronomy underwent a radical transformation in the 1990s with the publication of the popular book *Maya Cosmos*, written by my late friend and colleague Linda Schele and her co-authors, archaeologist David Freidel and writer Joy Parker. Much of the book tells the sometimes very personal story of Schele's process of discovery, as she came upon a new way of looking at Maya art and

---

[*] See Aldana, 2005. For further discussions of complex issues surrounding the "Venus War" glyph complex, see Aveni and Hotaling, 1994; Nahm, 1994; and Chinchilla Mazariegos, 2006.

iconography, especially those mythological scenes relating to Creation on 4 Ahaw 8 Kumk'u. In often breathless prose, she wrote of how she came to realize that numerous motifs in the iconography could be read as parts of figurative star maps, depicting the various elements of the night sky in orientation with one another. Numerous icons of Maya art, in her analysis, became specific elements of such maps, placed in relation to one another as depictions of the stars.

Many of Schele's theories have to do with the mythology of Creation associated with the year 3114 BC. We'll recall that mythological texts from the Classic period, such as Quiriguá Stela C, describe a location for the Creation event as being a divine hearth: three stones set up by a series of gods on August 11 or 13 of that year (depending on the correlation used). The name of the location was the "New Hearth Place," suggesting fairly clearly that the principal event of that time was the renewal or lighting of a sacred fire. Schele was intrigued to learn that the K'iché' Maya of Momostenango, in highland Guatemala, identify a triangular arrangement of stars in Orion as a celestial hearth. As she herself wrote, "These [stars] had to be the same three stones that were laid at Creation,"[8] and

> we deduced that these three stones of Creation are symbolic prototypes for the hearthstones used in Maya houses for over three millennia. As the hearthstones surround the cooking fire and establish the center of the home, so the three stone thrones of Creation centered the cosmos and allowed the sky to be lifted from the Primordial sea.[9]

The Maya Creation narrative now had an overt astronomical aspect to it, where the participants and things associated with 4 Ahaw 8 Kumk'u could be interpreted as stars, constellations, or other heavenly bodies. Schele also interpreted one enigmatic passage from a text at Palenque as an indication that this same Creation episode involved the "raising" of a world tree in the northern sky, and a "turning" of the night sky around a northerly pivot point, today associated with Polaris. As it turns out, both interpretations are based on highly problematic readings of the glyphs.

Schele's vision of a profound astronomical component within these mythical texts led her to approach Maya symbolism in a radical new way. According to her and others who worked closely with her at the time, much of the imagery on vase paintings and stone tablets could be interpreted with an astronomical template, using star maps of the sky as guides for understanding the works' compositions. For example, many of the common sky bands or serpents in Maya iconography represent the ecliptic, that is, the plane of the solar system that, from earth, is seen as the "path" in the sky wherein the sun, moon, and planets all perform their movements over the course of the day and night.

A good example of Schele's approach comes from her analysis of the famous "Blowgunner Vase," now in the Museum of Fine Arts in Boston, with its image of the hunter Hunahpu shooting a bird from the naturalistic-looking tree in front of him. This is a true masterpiece of Native American art, relating a key episode in the origin mythology described in the Popol Vuh. For Schele, however, this image was far more than a depiction of a scene from ancient folklore or mythology. For one thing, the image of the large scorpion—itself a masterwork of calligraphic painting—led her to consider an astronomical interpretation for the entire scene. Other scorpion representations in the art, some associated with sky bands and star symbols, had caught the at-

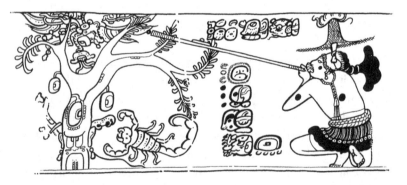

Mythological scene from the Blowgunner Vase, a Classic Maya ritual vessel from the eighth century AD. (Drawing by M. Zender)

tention of earlier scholars, who, positing a remarkable coincidence with the old world, considered it a possible correspondence to the zodiacal sign Scorpio. The point is still debated, but Schele was prepared to side with the advocates of the correspondence. For her this was the key to understanding the whole image. It led her to consider that the whole scene was a stellar map, representing the entire arc of the Milky Way (the tree) spanning from north to south, with the Big Dipper at one end (the Principal Bird Deity at the treetop) and the constellation Scorpio near its base, in the southern sky.

Schele took this insight and applied it to a wide range of other images, including the famous sarcophagus lid from Palenque, with its posthumous portrait of Palenque's king K'inich Janab Pakal. The central cross-shaped image of the scene is a stylized tree, with the Principal Bird Deity again shown perched above. This, by analogy, had to be the Milky Way, with the Big Dipper near its top, or northern, extension. The two-headed snake draped over the middle of the tree was, therefore, the ecliptic plane, the "horizon" of the solar system along which we perceive the movement of the planets and the sun. The deceased king in Schele's analysis is falling into the southern horizon after his death or, as the Maya scribes of Palenque put it, "entering the road." Schele approached numerous other mythological scenes in a similar way, realizing that "every major image of Maya cosmic symbolism was probably a map of the sky."[10]

She presented her new and exciting theories to crowds of scholars and enthusiasts at the 1993 Maya Meetings in Austin, Texas. There was a great deal of excitement in the audience as she shared her infectious enthusiasm with everyone, telling of her many "eureka moments" as she transformed her understanding of how Maya art and iconography worked. Many colleagues agreed with her—it was often hard to argue with Linda given her outright love and enthusiasm for the subject—while others, including me, remained unconvinced of many details. To me, some of her identifications were entirely plausible, but many others seemed too much like assumption piled on assumption. For example, in one analysis the Milky Way was identified as the cosmic tree; in another scene, as a canoe bearing gods into the Underworld.

It all seemed too loose and messy, even if she was able to pull together a narrative to explain a number of images using her new ideas. I don't think Linda ever quite forgave me for not embracing her ideas more, even though the interpretations in the pages of *Maya Cosmos* gave all Mayanists great food for thought, and still do. Unfortunately her untimely death several years later never allowed for these star map ideas to become as fully developed as she perhaps would have liked following her initial deluge of excitement in the early 1990s. *Maya Cosmos* presents Schele's astronomical theories in raw and unprocessed form, and perhaps without the necessary reflection that those ideas deserved.

Now that several years have passed, I can say that very few of the astronomical interpretations in *Maya Cosmos* have stood up well to close scrutiny. Interpreting Maya iconography requires a certain amount of prooflike argument, akin to deciphering hieroglyphs, where testing and retesting are always important. For me, the Blowgunner Vase need not be seen as a star map, but rather as a scene of an important myth, much as previously supposed. Is the scorpion Scorpio? Maybe. But what if it isn't? Might the original myth of Hunahpu shooting his blowgun have incorporated a scorpion as part of a story now lost to us? Can't we take it more at face value, as a depiction of an actual scene, with a blowgunner, a tree, and a scorpion? Other images of Hunahpu with his blowgun do not include a scorpion, so what does that tell us about the image's possible meaning? Similar doubts come to mind when considering and evaluating nearly every iconographic interpretation presented in *Maya Cosmos*. And as difficult as it is to say, I feel there's good reason to reject many of the assumptions underlying the book's astronomical model of Maya iconography.

So where does this leave us today in the study of the ancient Maya and their religious art? It's inevitable that calendrics and astronomy will continue to occupy an important place in our vision of ancient Maya culture and civilization, but not in the same defining way they have almost since the beginning of research on the subject. Looking at records of planets will forever remain a key aspect of studying the ancient codices, for instance, and certain solar and astral alignments will always be important to an understanding of particular aspects of, say, architectural design. But as we learn more and more about who the Maya were and what they wrote about, I doubt astronomy can still hold enough weight to be the powerful, overarching paradigm for the culture that it once was. For decades now the notion of the "stargazing Maya" has been a forceful and influential one, not just in the popular imagination, but in academic circles as well. Astronomy was important, but no more so than elsewhere in Mesoamerica. In future years, I suspect further decipherments and archaeological discoveries will balance this long-standing intellectual tendency with a more realistic and earthly vision of the ancient Maya as real people with a real history.

Maya of the New Age

We've touched on early explorers and writers saw ancient Mesoamerican peoples as historical anomalies: "Indians" who didn't quite fit the nineteenth-century ideal of "Indian" and the "noble savage." They were somehow outsiders to human history as it was then understood, and the stage was soon set for romanticized images of the Maya as exotic and mysterious, as the quintessential "lost" civilization. Our understanding of the ancient Maya especially has long wrestled with this intellectual baggage. A century ago, as research zeroed in on their keen awareness of astronomy and numerology, the Maya became exoticized in new ways, as great intellectuals of the New World. The subsequent wrenching of the Maya into the world of history and violent warfare was jarring, and I think many of us still strive in our own way to reconcile such utterly disparate images of the culture. In his 2006 film *Apocalypto*, Mel Gibson did so by creating two Maya cultures—one warlike, one peaceful—each unaware of the other! In much more serious academic studies, as we've seen, the reconciliation of two ideals has come about more through attempts to understand the newly deciphered contents of Maya history and mythology in terms of astronomy—usually, I find, without much success.

Which brings us to where we are now. Astronomy remains a vital subfield within Maya and Mesoamerican studies, although no longer at center stage the way it was for so long. Because now, in the last couple of decades, we are in a remarkably exciting time, when Maya writing can be read and largely understood. We can read hundreds of ancient historical and mythical texts, and look forward to the decades it will take to sort through the vast number of sources now available written in an authentic Maya voice. Numerous ancient cities have been excavated, and we have a good sense of what forces led to the rise and fall of Maya civilization over the span of two thousand years. The Maya are, for the first time since their discovery centuries ago, a people with a history.

I have to wonder, though, if the discovery of these new and very exciting windows into history and mythology, revealing the profound "normalcy" of the Maya as a human culture, has created something of a backlash. Old ideas die hard, and it strikes me that there are many who will continue to see the ancient Maya as deeply exotic and different. I'm convinced that much of the current 2012 phenomenon might be such a pushback, reasserting in the face of new discoveries that the Maya were unusually perceptive astronomers, that they somehow had a special and unique understanding of the workings of the sky and the cosmos. Today when I read New Age writings about the Maya and 2012, whether they be about their awareness of some change in "human consciousness" or about their knowledge of a "galactic alignment," it's clear that some attitudes haven't changed since those early, off-base descriptions of the Maya as ethereal folk always looking up at the night sky. Several New Age spiritual leaders have in different ways appropriated the Maya calendar, or, to be accurate, their poor, superficial understanding of it, in order to use it as a source of personal spiritual insight, inspiration—not to mention monetary income.

Where exactly did these ideas come from? Some of them can be traced back to an offhand remark written in 1966 by my esteemed colleague Michael Coe, professor emeritus of anthropology at Yale University. Coe has made numerous key contributions to Olmec and Maya archaeology and has written many popular and accessible books on Mesoamerican topics. Some forty years ago he speculated in one such book that the upcoming end of the thirteenth bak'tun might have been thought of an as "Armageddon" that would see the destruction of the world. In making this passing claim, he was drawing in a reasonable way upon some of the familiar concepts of mythic history from elsewhere in Mesoamerica, most notably upon the Aztecs' cycles of world destruction and creation. As he admits, then as now, no Maya source makes any such claim about 2012; he was simply speculating about the ways in which the ancient Maya could have conceived of that important day in their calendar.

Yet over the ensuing years the 2012 doomsday idea took hold, at least in some quarters, especially among those who embraced a certain brand of Western American mysticism that intersected with the counterculture movement of the 1960s. One proponent was a man named Frank Waters, known as the "Grandfather of Southwestern Literature." Based in later life in Taos, New Mexico, Waters wrote books that included a number of works on Indian mysticism, including perhaps his most famous work, *The Book of the Hopi*. He wasn't an anthropologist, however, and his writings on Hopi religion, myth, and cosmology, while of significant interest, were highly unconventional, and based on no firsthand knowledge of the Hopi language.* In 1971 he published a work called *Mexico Mystique: The Coming Sixth World of Consciousness*, where a good many of the current threads of 2012 thinking can be traced.

Waters was a big-picture sort of thinker: he evokes a Jungian idea of humanity's collective unconscious to assert that ancient Mesoamerican deities reflect a "universal meaning [that] is as pertinent now as it was two thousand years ago." He was keenly interested in unifying the many diverse threads of Southwestern and Mesoamerican mythology and religion into a cohesive whole, almost as if he were defining for himself and others the tenets of a new organized religion. So, for example, he completely fused Aztec and Maya belief in suggesting that the Creation date 13.0.0.0.0 4 Ahaw 8 Kumk'u marked the beginning of "the Fifth Sun." And from this, he asserted that the close of the Mayas'

---

* See Sitler, 2006, for how New Age thinking misrepresents traditional Mayan religion. See Geertz, 1996, for a general overview of the New Age appropriation of Hopi philosophy, a situation that bears a striking resemblance to the case of the Maya.

thirteen-bak'tun "Great Cycle" in 2011 or 2012 (he had the date slightly off) would mean the world's end, destruction of the world by earthquakes, which he based on the catastrophism we have seen related in the Aztec "Legend of the Suns." Aztec, Maya, and Hopi religions were, to Waters, all of a piece, what he called "Nahuatl-Maya myth." To a certain extent he wasn't completely off base in this assertion—they are, after all, culturally and historically related. But more care was called for when proposing a unified system of mythology and a cohesive religious narrative. No such thing ever existed in ancient Mesoamerica. Ultimately, it was Waters's collapsing of two different mythical narratives that helped light the spark of the 2012 movement.

Waters's *Mexico Mystique* was a direct and acknowledged influence on one of the most influential of the New Age voices on the Maya calendar and the 2012 movement, José Argüelles, author of the spectacularly incomprehensible yet popular treatise *The Mayan Factor: Path Beyond Technology*, first published in 1987. Argüelles claimed to be the "Messenger of the Command of Pacal Votan," and in that important-sounding role, he makes claims that the Maya were "galactic agents" who'd come to earth in order to "place earth and its solar system in synchronization with the larger galactic community." The 2012 date for him "bodes nothing less than a major evolutionary upgrading of the light-life—*radiogenetic*—process which our planet represents." Today, Argüelles heads something called the "Galactic Research Institute," which gives us some idea of where his thinking still lies. (As you might imagine, the institute has nothing to do with real astronomical study or research.)

Argüelles is perhaps best known for his advancing the idea of the so-called "Harmonic Convergence," which was supposed to occur back in 1987, based on his rather odd understanding of the Maya calendar. That year saw a supposedly meaningful alignment of a number of planets with the solar system, which, to Argüelles, signified the ushering in of a new era energized by world peace and harmony, and a rejection of war and strife. (I'm not so sure it met with much success, given the current events of these last few years.) The language used to describe that spiritual event sounds remarkably like what we hear today in connection with the year 2012.

By now it ought to be clear that in terms of erudition and academic grounding, *The Mayan Factor* is utter nonsense. Argüelles seems perfectly serious in his claims that "galactic" Maya culture is not of this earth. He consciously fabricated his ideas out of the creative stream of his own consciousness, and admits as much in many of the passages of his books. Needless to say, the point of his work was not to represent Maya culture in any dispassionate way, but to convey something of his own self, using the ancient Maya as a crude backdrop, with a vocabulary of images and concepts that illustrate what amounts to the author's own personal journey of the mind and mystic revelation. The "truth" the Maya seek is an inner, spiritual one, he believes, oriented toward personal growth and awareness. In this way, Argüelles follows in the genre of Carlos Castañeda and similar self-promoting philosophical voices of the counterculture era. "Authentic" representations of Maya culture and history are hardly a concern in such works, so such writers might well find some encouragement in my skepticism, given that I'm a member of the academic establishment, or, as Argüelles himself would dismissively say, a "materialistic archaeologist."

Nowadays there are many gurus who claim to know something about 2012 and its meaning. One well-known member of this club is writer and psychedelic Daniel Pinchbeck, author of *2012: The Return of Quetzalcoatl.* His ideas are meandering and difficult to comprehend, and perhaps the most radical in terms of having any basis in Maya religion or culture. For Pinchbeck, 2012 is a "transformative threshold on our world," a time when people must embrace "indigenous shamanic knowledge" to ensure human survival and awareness in a time of world crisis. In essence, his very personal writings are about our world and the great problems it faces, and how "Mayan prophecies" of the end of a cycle present all of us with an opportunity to metamorphose and set things straight. How did Pinchbeck find his own shamanic, inspired path to understanding 2012? Partially through a hallucinogen-induced vision of "Quetzalcoatl" during a visit by Pinchbeck to an Amazonian tribe. (By the way, Quetzalcoatl was more a central Mexican deity, not so much Maya.) Such self-centered writings are a bit hard to take, I must say, but they do appeal to many who seek a broader philosophical context for the economic, political, and environmental problems our world now confronts. That may be fine, but it's important for all to realize that Pinchbeck's claims about Quetzalcoatl's "return" in 2012 don't really have any basis in authentic Mesoamerican culture or philosophy.[11]

Many other New Age or pseudo-science types claim similar insight into and understanding of what 2012 will mean, Carl Johann Calleman and Lawrence E. Joseph among them.[12] Tellingly, none of these pseudo-scholars or New Age authors who purport to explain or understand 2012 seems to agree with the others. As with Pinchbeck, each of their "revelations" is self-proclaimed and internalized, conveniently detached from any scientific or verifiable evaluations. Of course this shouldn't be too surprising, given that science and cultural knowledge are far from these authors' immediate concerns.

These days a prominent voice in the New Age interest in 2012 is the writer and "independent researcher" John Major Jenkins, author of *Maya Cosmogenesis 2012* and *2012: The Real Story*, among other works. Although he's been described as an anthropologist, Jenkins is not an academic, and his most publicized ideas about astronomy fall well outside the mainstream of Mesoamerican scholarship. For him, the turn of the cycle means the dawn of a new age, when "our basic assumptions and foundational values will be exposed, and we will have the opportunity to embrace values long since driven under the surface of our collective consciousness." The Maya predicted, in his view, a change from "one extreme form of social organization to another," with the turn of the baktun period offering humanity a "conscious relationship with each other and a creative participation with the Earth-process that gives birth to our higher selves."[13]

Despite such vague declarations, Jenkins's claims are far more grounded in reality than those of Argüelles. How could they not be? He mostly focuses on the astronomical symbolism of carvings and architectural layouts at the important ruins of Izapa, in southern Mexico, to make a complex case that the cycle ending in 2012 has a real astronomical basis. Because of its latitude, Izapa, a Preclassic Maya center, has long been a place of interest for those attempting to explain the 260-day calendar, but Jenkins has gone far beyond conventional wisdom to claim that the monuments at the site "explicate the 2012 prophecy and the spiritual teachings that apply to cycle endings." To him, the "2012 end-date is firmly established and is a true and accurate artifact of the Mayan philosophy of time." His specific claim is that on December 21, 2012, the sun will rise in direct alignment with the center of the Milky Way galaxy, and that the Maya somehow were aware of this future occurrence. Repeating ideas first made by Frank Waters, Jenkins also claims that 2012 is the end point of a complete precession cycle (twenty-six thousand years), of which the Maya, again, apparently had some knowledge and which they were able to predict.

Although Jenkins illustrates his books and articles with star charts and detailed interpretations of Izapan sculptures, lending all of it the appearance of scholarship by someone who knows his stuff, nearly all of it is wrong. There are also serious problems with his claims about the astronomy of December 21, 2012, not the least of which is that the Maya could not have had any idea of a galactic center in the first place. The center of the Milky Way was recognized by modern science only a few decades ago, through the methods developed in radio astronomy. Aside from that, Jenkins's ideas simply aren't informed by any deep knowledge of Maya religion or art. He was clearly greatly influenced by *Maya Cosmos* and Linda Schele's method of using astronomical maps to interpret iconographic scenes. These, as we've seen, were quite controversial when they were first published and don't hold up well today. Jenkins's fanciful interpretations of the imagery on early Maya stone monuments at Izapa are based on some of the same general principles, but with far less insight and grounding in Maya iconography and religious culture. I'm sure that Schele wouldn't have agreed with any of his interpretations had she lived to see the recent 2012 brouhaha. Oddly, Jenkins has conveniently claimed that interpreting "Izapan iconography is too important to be left to the iconographers." No doubt such a statement would have rankled Schele, the premier Maya iconographer of her time, as much as it bothers me. It's a bit like saying that "physics is too important to be left to the physicists."

Revealingly, the writings of "2012-ologists" often take on a very personal tone. Jenkins describes his thinking on Maya time and cosmology as his own "revelation" or "journey" along a "deepening initiatory path."[14] As with many others who have an interest in what the turn of the bak'tun cycle "means," the stakes for Jenkins appear very personal, even inward looking. This, I suspect, makes debating his particular ideas and especially those of the more radical New Age thinkers well nigh impossible. Once the discussion turns to our own time and our own "selves," it is very hard to argue particulars of Maya art and astronomy. I've become increasingly aware that these writers and "independent researchers" are not so much interested in what the Maya actually were trying to communicate, but rather are after something that's spiritually based, and are looking forward to changes in our modern, chaotic world. I wish them all well in their personal quests, but it's important to stress that the substance of their ideas has little if anything to do with the ancient Maya and the messages they were communicating in their art and writings.

The Tortuguero Tablet

I've avoided saying much about my own interpretation of the Maya "end date" of 2012, because the reality of the ancient Maya calendar and cosmology seems endlessly more fascinating and authentic than anything so far penned by a guru or New Age "expert" on Maya timekeeping. Now, allow me to be as unambiguous as I possibly can: *no authentic Maya text foretells the end of the world in 2012, or of any destructive event happening in connection with the turn of the thirteenth bak'tun.* There will no doubt be important events in national and world history, and maybe a fair number of unfortunate tragedies and disasters. But the world we live in will not come to a screeching halt. (Anyone reading these words in 2013, 2020, or some year farther on after "doomsday" can definitely back me up on this point.) And because the Maya never once claimed in any ancient or traditional source anything of the kind, those who say otherwise have no idea of the Maya calendar's real structure or significance. Put simply, they have no idea what they are talking about.

The change in the bak'tun is of course a real and noteworthy feature of the Long Count calendar, and if we assume our correlation is true, then the day December 21 or 23, 2012, will see the shift of the bak'tun number from 12 to 13, from 12.19.19.17.19 to 13.0.0.0.0. This is clearly a key station in the structure of the calendar, given the great importance the Maya placed on the number 13 in their overall cosmology and conception of time. And as we've seen, the 2012 date represents a numerical recurrence, an anniversary of sorts, of the full Long Count date of the Creation base date 4 Ahaw 8 Kumk'u in 3114 BC. That is where its importance lies: not as an "end" by itself, but as the first of many future repetitions.

We can illustrate it in this way, using the mind-numbing quantities of time as expressed in the Grand Long Count described in the previous chapter. The upper date is the "Creation" of 3114 BC, and the second is the repetition that is set to occur in 2012.

13.13.13.13.13.13.13.13.13.13.13.13.13.13.13.13.13.13.13.0.0.0.0
4 Ahaw 8 Kumk'u
+13.0.0.0.0
13.13.13.13.13.13.13.13.13.13.13.13.13.13.13.13.13.13.13.0.0.0.0
4 Ahaw 3 K'ank'in

It's important to realize that an identical repetition of this same Grand Long Count station will take place in the distant future in a little more than one hundred thousand years, after thirteen Piktuns have elapsed after 2012. The upcoming anniversaries in 2012 and beyond serve as a reflection of the Mayas' deeply felt concern for and interest in a highly structured cosmology, with time as one of its chief dimensions.

Only one ancient inscription makes clear reference to 2012. It's a partial monument from the site named Tortuguero, located in Tabasco, Mexico, not far from the much more famous ruins of Palenque. Tortugero hardly exists today as an archaeological site. Decades ago its temples were wiped off the map, their stones removed for use as gravel in construction of a nearby highway; today a large gravel factory oc-

A part of the Tortuguero tablet (Monument 6), with its reference to the 2012 bak'tun ending (13.0.0.0.0 4 Ahaw 3 K'ank'in) in the last two columns of glyphs. (Drawing by the author)

cupies its original setting. In the course of Tortuguero's destruction in the 1960s, a number of important sculptures came to light, none ever seen in its original setting by an archaeologist. Pieces were sold off to art collectors, and others made their way to the museum in the neighboring city Villahermosa.

Of all the Tortuguero sculptures, the most imposing is a beautiful limestone slab we call Monument 6. This large stone probably graced the interior wall of a temple, now destroyed, describing the shrine's ritual significance. The monument is now in several fragments, with some parts still missing, as it's currently displayed in the museum. One unseen portion made its way into the art market and was sold into private hands in the United States. While its current location remains unknown, photographs of it do exist, and allow us to reconstruct most of the monument's original form. It has an odd *T* shape, although the upper "wing" of its left side is missing. Perhaps it, too, is in a private collection, waiting to be seen and studied by scholars.

The Tortuguero monument displays a very long and mostly well preserved hieroglyphic inscription. Following the format of most Maya texts, it shows an array of squares arranged in a grid, each portion holding a set of complex signs, all rich in visual detail. The words encoded in these glyphs convey events in the life of an important local king named Bahlam Ajaw (Jaguar Lord), who lived in the early decades of the seventh century AD. He was the king, or *k'uhul ajaw,* who ruled over the court called Baakal, which historians usually associate with the nearby center of Palenque; kings there, such as the famous K'inich Janab Pakal, were curiously also "holy lords" of Baakal. Based on the historical records we know from both Tortuguero and Palenque, we suspect that Tortuguero was a temporary seat of the Baakal royal court, perhaps moved there during a time when the larger center of Palenque was suffering military defeat at the hands of the aggressive Kan kingdom, based far to the east. Bahlam Ajaw may have been a relative of K'inich Janab Pakal and others at Palenque, but we can't be sure.

Bahlam Ajaw became king on February 4, AD 644, and in a few years he engaged in a number of military conquests of his own, including a defeat of the nearby city of Comalcalco in 649.* Twenty-five years into his rule, on January 11, 669, he dedicated a temple or shrine in the community over which he ruled. This place is now gone, destroyed when the ruins were razed in the 1960s. Monument 6 was from that building. Bahlam Ajaw evidently performed this dedication as a refurbishment or rebuilding of an older structure mentioned in the tablet, which had been built by an ancestor in 510. This shrine, called a *pibnaah,* or "oven house," according to the text, must have been of profound importance. Perhaps it was meant to house the effigy of a significant patron god, or maybe it was dedicated to some ancient ruler as a funerary temple. Because of the depredation of the ruins and the loss of their original context, we will never know. But Bahlam Ajaw thought highly enough of his deeds to link the dedication of his building not only to a distant ancestor's work, but also to an event over a thousand years in the future. Part of the very final passage says this: "In two days, nine-score days, three years, eight score years, and three times four hundred years, the thirteenth bak'tun will end, [and] 4 Ahaw, 3 K'ank'in will happen." In a wonderfully ironic twist, a large crack runs through the last few glyphs of the inscription, where once there may have been some statement about 2012 and its significance.†

Only one readable glyph remains well preserved in this tantalizing final passage: the name of a poorly understood Maya god, Bolon Yokte' K'uh, "Nine Pole God," or perhaps "Nine Pole Gods" (it might

* Our knowledge of the ruler Bahlam Ajaw and his history comes from several Tortuguero inscriptions. See Gronemeyer, 2006, for a detailed overview of the site's monuments and hieroglyphic texts.

† My colleague Stephen Houston makes a compelling argument that these final damaged hieroglyphs may not describe 2012 at all, but instead pertain to a restatement of the ancient building dedication featured earlier in this same inscription. See Houston, 2009. Such temporal leaps forward and backward are known in other Mayan texts, as Houston notes. If he's right, then we're left with no authentic Mayan record *anywhere* about the significance of the 13.0.0.0.0 4 Ajaw 3 K'ank'in date. I remain undecided on the matter, seeing both temporal placements of the final passage as possible; I suspect that the damage to the text presents too many ambiguities for the issue to be resolved to everyone's satisfaction.

be a single or collective reference). He was one of the characters or groups of deities we saw mentioned on the famous Vase of the Seven Gods as being "ordered" or "arranged" on the day of Creation, in 3114 BC. Some claim that Bolon Yokte' K'uh (or Bolon Yokte') was a war deity, but there's no strong evidence of this. Even though we cannot read the Tortuguero passage with much ease, and may never do so, I find it fascinating that Bolon Yokte' K'uh, whoever or whatever he was, might be making some sort of reappearance on the anniversary of the day he was placed in proper relation to other gods of the cosmos.

Back in 2006, I was asked my opinion on the Tortuguero tablet and its mention of 2012. I offered a quick analysis and posted it on a Mesoamerican studies listserv, mentioning my then very tentative reading of a glyph next to the name of Bolon Yokte' K'uh as "his descent." Within weeks I was shocked to see websites across the world referring to "David Stuart's analysis" of the Tortuguero prophecy as "Bolon Yokte' K'uh will descend." Many New Age writers have taken this obscure and imperfectly understood reference and run with it, creating a whole new mythology that I'm sure we'll be hearing more about as 2012 approaches. One major problem with this, though, is that the "descent" reading was probably wrong to begin with. The entire last few glyphs are either too broken to be read or simply too difficult for modern epigraphers to analyze with much confidence. One thing is clear to me, at least: nowhere does it say that Bolon Yokte' K'uh will descend come the end of the bak'tun.

So that's it. The Tortuguero tablet mentions the end of bak'tun thirteen in 2012 just as if it were any other predictable period ending. The date was included in the long text as a rhetorical device anchoring the history of the local ruler Bahlam Ajaw in proper "big picture" context, juxtaposed with a date far in the past of his own dynastic history. The citation of 2012 gave symmetry and balance to that narrative, with Bahlam Ajaw's accomplishments in the chronological center. Does the Tortuguero passage say anything meaningful about what will actually happen in 2012? Absolutely not. Nor does any other ancient Maya text, for that matter.

◇

I'm sympathetic to the many people who prefer to look to traditional native beliefs for their own spiritual awareness. But I can't lend much sympathy to those who so patently misrepresent ancient Maya belief and culture in their own varied efforts to inspire or forge some new spiritual benefit. Maya "mysticism" need not be made up whole cloth from scanty evidence, now that after decades we can finally read ancient religious texts and ponder their meanings. So, assuming, as I do, that authenticity of culture and heritage is still important and carries meaning for the people of today, I'll stick my neck out and hereby assert that the ancient Maya calendar has never truly been understood for what it actually was, and is. Centuries ago it was a structure upon which was built a profoundly complex religious world and cosmology, intimately tied to the ideology of kings and nobles. It was the framework for the understanding of history and its "folds" in the distant past, and for projecting established patterns into the future. Today, the surviving day counts in some Guatemalan communities serve as interpretive tools for the routine lives of people, many of them isolated and struggling to adapt to the modern world. The history of the calendar stands in stark contrast to today's pervasive idea that Maya time somehow holds relevance for all of us, offering a greater understanding of human history or of the world's fate in the very near future. It actually doesn't. I won't speak for others in the field of Maya archaeology and research, but I personally have a hard time understanding how and why the ancient calendar should have any connection to the lives of anyone other than the Maya themselves, who, along with a scattering of other native Mesoamericans, have closely guarded their tradition of timekeeping for centuries since the Conquest. For the Maya, both ancient and modern, the interwoven structures of time did what calendars "do" for countless cultures throughout human history: they gave significant meaning to their lives, politics, and rituals, along with simply giving order to the passing of the days. In this profoundly human way, the "mysterious" Maya are really not so mysterious after all.

APPENDICES

APPENDIX 1 Comparison of the Twenty Day Names in Four Mesoamerican Languages

ZAPOTEC

Chilla, Chiylla	Crocodile
La, Quiy, Guiy	Live Coal, Fire, Wind(?)
Guela, Ela	Night
Gueche, Quichi, Achi	Frog, Iguana
Zee, Ziy, Cee, Ziye	Misfortune, Serpent, Young Corn
Lana, Laana	Soot, Rabbit
China, Chiyña	Deer
Lapa, Laba	Divide, Crown, Garland
Niça, Queça	Water
Tela, Tella	Facedown, Dog
Loo, Goloo	Monkey(?)
Pija	Twisted(?)
Quij, Laa, Nij	Reed
Gueche, Eche, Ache	Jaguar
Naa, Na, Ñaa	Mother(?)
Guiloo, Loo	Raven(?), Eye(?)

Xoo	Earthquake
Opa, Gopa, Oppa	Vapor from Earth
Appe, Ape	Cloudy
Lao, Loo	Face

MIXE°

Hukpii	Root
Shra'a	Wind
Jou	Palm
Juun	Hard, Solid, Resistant
Tsa'an	Serpent
Uj	Earth, World
Koy	Rabbit
Naan	Deer
Nu'un	Water, River
Jo'o	Vine, Dog
Jaymi	Fine White Ashes
Tuuta	Tooth
Kapy	Reed
Kaa	Jaguar
Juiky	Tobacco
Pa'	Edge, Border, Side
Ujshi	Earthquake
Tap	Covered up, Blackening, Darkening
Muy	Grass
Jugwin	Bee, Fontanelle, Point, Eye

YUKATEK MAYAN†

Imix	Water spirit(?)
Ik'	Wind, Breath
Ak'bal	Night, Darkness

° Mixe day names are based on Caso, 1967; and Lipp, 1991, p. 63.

† From the sixteenth century; meanings based on classic visual forms.

K'an	Ripe Maize(?)
Chikchan	Snake
Kimi	Death
Manik'	Deer(?)
Lamat	Star
Muluk	Water jar
Ok	Dog
Chuwen	Monkey Deity
Eb	Tooth
Been	Reed(?)
Ix	Jaguar
Men	Eagle
Kib	?
Kaban	Earth
Etz'nab	Knife
Kawak	Storm
Ahaw	Ruler

AZTEC (NAHUATL)

Cipactli	Alligator
Ehecatl	Wind
Calli	House
Cuetzpallin	Lizard
Coatl	Snake
Mizquitl	Death
Mazatl	Deer
Tochtli	Rabbit
Atl	Water
Itzcuintli	Dog
Ozomatli	Monkey
Malinalli	Twisted
Acatl	Reed
Ocelotl	Jaguar
Cuauhtli	Eagle

Cozcacuauhtli	Vulture
Ollin	Earthquake
Tecpatl	Flint-knife
Quiauitl	Rain
Xochitl	Flower

APPENDIX 2 The Day Names in Three Mayan Languages

YUKATEK [º]	TZELTAL [†]	K'ICHE [‡]
Imix	*Imox*	*Imox*
Ik'	*Ik'*	*Iq'*
Ak'bal	*Wotan*	*Aq'abal*
K'an	*K'anan*	*K'at*
Chikchan	*Abak'*	*Kan*
Kimi	*Tox*	*Came*
Manik'	*Moxik*	*Kih*
Lamat	*Lambat*	*Q'anil*
Muluk	*Mulu*	*Toj*
Ok	*Elab*	*Tz'i*
Chuwen	*Batz'*	*Batz'*
Eb	*Ewob*	*E*
Ben	*Been*	*Aj*
Ix	*Hix*	*Ix*
Men	*Tz'ikin*	*Tz'ikin*
Kib	*Chabin*	*Ajmak*
Kaban	*Chik*	*Noh*
Etz'nab	*Chinax*	*Tihax*
Kawak	*Kahok*	*Kahuk*
Ahaw	*Ajual*	*Junajpu*

[º] From the sixteenth century.

[†] From the eighteenth century.

[‡] Modern.

APPENDIX 3 Mayan Month Names from Yucatán and the Classic Period

YUKATEK	CLASSIC MAYAN
Pop	*K'anjalaw*
Woh	*Chakat / Wooh*
Sip	*Ik'at*
Sotz'	*Suutz'*
Tzek	*Kasew*
Xul	*Tzikin*
Yaxk'in	*Yaxk'in*
Mol	*Mol*
Ch'en	*Ik'sihoom*
Yax	*Yaxsihoom*
Sak	*Saksihoom*
Keh	*Chaksihoom*
Mak	*Mak*
K'ank'in	*Uniw/K'ank'in*
Muan	*Muwaan*
Pax	*Paax*
K'ayab	*K'anasiiy*
Kumk'u	*Hulohl(?)*
Wayeb	*Ti'wayhaab(?)* (final five-day period)

APPENDIX 4 Historical Period Endings of Cycles 8, 9, and 10 (AD 297–909)

LONG COUNT	CALENDAR ROUND	GREGORIAN EQUIVALENT[*]
8.13.0.0.0	9 Ahaw 3 Sak	December 12, AD 297
8.13.5.0.0	2 Ahaw 13 Ch'en	November 16, 302

[*] 584,283 GMT correlation constant

LONG COUNT	CALENDAR ROUND	GREGORIAN EQUIVALENT
8.13.10.0.0	8 Ahaw 13 Mol	October 21, 307
8.13.15.0.0	1 Ahaw 8 Yaxk'in	September 24, 312
8.14.0.0.0	7 Ahaw 3 Xul	August 29, 317
8.14.5.0.0	13 Ahaw 18 Sotz'	August 3, 322
8.14.10.0.0	6 Ahaw 13 Zip	July 8, 327
8.14.15.0.0	12 Ahaw 8 Woh	June 11, 332
8.15.0.0.0	5 Ahaw 3 Pop	May 17, 337
8.15.5.0.0	11 Ahaw 3 Kumk'u	April 21, 342
8.15.10.0.0	4 Ahaw 18 Pax	March 25, 347
8.15.15.0.0	10 Ahaw 13 Muwan	February 27, 352
8.16.0.0.0	3 Ahaw 8 K'ank'in	January 31, 357
8.16.5.0.0	9 Ahaw 3 Mak	January 5, 362
8.16.10.0.0	2 Ahaw 18 Sak	December 11, 366
8.16.15.0.0	8 Ahaw 13 Yax	November 14, 371
8.17.0.0.0	1 Ahaw 8 Ch'en	October 18, 376
8.17.5.0.0	7 Ahaw 3 Mol	September 22, 381
8.17.10.0.0	13 Ahaw 18 Xul	August 27, 386
8.17.15.0.0	6 Ahaw 18 Tzek	August 1, 391
8.18.0.0.0	12 Ahaw 8 Sotz'	July 5, 396
8.18.5.0.0	5 Ahaw 3 Sip	June 9, 401
8.18.10.0.0	11 Ahaw 18 Pop	May 14, 406
8.18.15.0.0	4 Ahaw 18 Kumk'u	April 18, 411
8.19.0.0.0	10 Ahaw 13 K'ayab	March 22, 416
8.19.5.0.0	3 Ahaw 8 Pax	February 24, 421
8.19.10.0.0	9 Ahaw 3 Muwan	January 29, 426
8.19.15.0.0	2 Ahaw 18 Mak	January 3, 431
9.0.0.0.0	8 Ahaw 13 Keh	December 8, 435
9.0.5.0.0	1 Ahaw 8 Sak	November 11, 440
9.0.10.0.0	7 Ahaw 3 Yax	October 16, 445
9.0.15.0.0	13 Ahaw 18 Mol	September 21, 450
9.1.0.0.0	6 Ahaw 13 Yaxk'in	August 25, 455

LONG COUNT	CALENDAR ROUND	GREGORIAN EQUIVALENT
9.1.5.0.0	12 Ahaw 8 Xul	July 29, 460
9.1.10.0.0	5 Ahaw 3 Tzek	July 3, 465
9.1.15.0.0	11 Ahaw 18 Zip	June 7, 470
9.2.0.0.0	4 Ahaw 13 Woh	May 13, 475
9.2.5.0.0	10 Ahaw 8 Pop	April 15, 480
9.2.10.0.0	3 Ahaw 8 Kumk'u	March 21, 485
9.2.15.0.0	9 Ahaw 3 K'ayab	February 23, 490
9.3.0.0.0	2 Ahaw 18 Muwan	January 27, 495
9.3.5.0.0	8 Ahaw 13 K'ank'in	January 1, 500
9.3.10.0.0	1 Ahaw 8 Mak	December 5, 504
9.3.15.0.0	7 Ahaw 3 Yax	November 9, 509
9.4.0.0.0	13 Ahaw 18 Yax	October 14, 514
9.4.5.0.0	6 Ahaw 13 Ch'en	September 18, 519
9.4.10.0.0	12 Ahaw 8 Mol	August 22, 524
9.4.15.0.0	5 Ahaw 3 Yaxk'in	July 27, 529
9.5.0.0.0	11 Ahaw 18 Tzek	July 1, 534
9.5.5.0.0	4 Ahaw 13 Sotz'	June 5, 539
9.5.10.0.0	10 Ahaw 8 Zip	May 9, 544
9.5.15.0.0	3 Ahaw 3 Woh	April 13, 549
9.6.0.0.0	9 Ahaw 3 Wayeb	March 18, 554
9.6.5.0.0	2 Ahaw 18 K'ayab	February 20, 559
9.6.10.0.0	8 Ahaw 13 Pax	January 25, 564
9.6.15.0.0	1 Ahaw 8 Muwan	December 29, 568
9.7.0.0.0	7 Ahaw 3 K'ank'in	December 3, 573
9.7.5.0.0	13 Ahaw 18 Keh	November 7, 578
9.7.10.0.0	6 Ahaw 13 Sak	October 12, 583
9.7.15.0.0	12 Ahaw 8 Yax	September 15, 588
9.8.0.0.0	5 Ahaw 3 Ch'en	August 20, 593
9.8.5.0.0	11 Ahaw 18 Yaxk'in	July 25, 598
9.8.10.0.0	4 Ahaw 13 Xul	June 29, 603
9.8.15.0.0	10 Ahaw 8 Tzek	June 2, 608

LONG COUNT	CALENDAR ROUND	GREGORIAN EQUIVALENT
9.9.0.0.0	3 Ahaw 3 Sotz'	May 7, 613
9.9.5.0.0	9 Ahaw 18 Woh	April 11, 618
9.9.10.0.0	2 Ahaw 13 Pop	March 16, 623
9.9.15.0.0	8 Ahaw 13 Kumk'u	February 18, 628
9.10.0.0.0	1 Ahaw 8 K'ayab	January 22, 633
9.10.5.0.0	7 Ahaw 3 Pax	December 27, 637
9.10.10.0.0	13 Ahaw 18 K'ank'in	December 1, 642
9.10.15.0.0	6 Ahaw 13 Mak	November 5, 647
9.11.0.0.0	12 Ahaw 8 Keh	October 9, 652
9.11.5.0.0	5 Ahaw 3 Sak	September 13, 657
9.11.10.0.0	11 Ahaw 18 Ch'en	August 18, 662
9.11.15.0.0	4 Ahaw 13 Mol	July 23, 667
9.12.0.0.0	10 Ahaw 8 Yaxk'in	June 26, 672
9.12.5.0.0	3 Ahaw 3 Xul	May 31, 677
9.12.10.0.0	9 Ahaw 18 Sotz'	May 5, 682
9.12.15.0.0	2 Ahaw 13 Sip	April 9, 687
9.13.0.0.0	8 Ahaw 8 Mol	March 13, 692
9.13.5.0.0	1 Ahaw 3 Pop	February 15, 697
9.13.10.0.0	7 Ahaw 3 Kumk'u	January 20, 702
9.13.15.0.0	13 Ahaw 18 Pax	December 25, 706
9.14.0.0.0	6 Ahaw 13 Muwan	November 29, 711
9.14.5.0.0	12 Ahaw 8 K'ank'in	November 2, 716
9.14.10.0.0	5 Ahaw 3 Mak	October 7, 721
9.14.15.0.0	11 Ahaw 18 Sak	September 11, 726
9.15.0.0.0	4 Ahaw 13 Yax	August 16, 731
9.15.5.0.0	10 Ahaw 8 Ch'en	July 20, 736
9.15.10.0.0	3 Ahaw 3 Mol	June 24, 741
9.15.15.0.0	9 Ahaw 18 Xul	May 29, 746
9.16.0.0.0	2 Ahaw 13 Tzek	May 3, 751
9.16.5.0.0	8 Ahaw 8 Sotz'	April 6, 756
9.16.10.0.0	1 Ahaw 3 Zip	March 10, 761

LONG COUNT	CALENDAR ROUND	GREGORIAN EQUIVALENT
9.16.15.0.0	7 Ahaw 18 Pop	February 13, 766
9.17.0.0.0	13 Ahaw 18 Kumk'u	January 18, 771
9.17.5.0.0	6 Ahaw 13 K'ayab	December 23, 775
9.17.10.0.0	12 Ahaw 8 Pax	November 26, 780
9.17.15.0.0	5 Ahaw 3 Muwan	October 31, 785
9.18.0.0.0	11 Ahaw 18 Mak	October 5, 790
9.18.5.0.0	4 Ahaw 13 Keh	September 9, 795
9.18.10.0.0	10 Ahaw 8 Sak	August 13, 800
9.18.15.0.0	3 Ahaw 3 Yax	July 18, 805
9.19.0.0.0	9 Ahaw 18 Mol	June 22, 810
9.19.5.0.0	2 Ahaw 13 Yaxk'in	May 25, 815
9.19.10.0.0	8 Ahaw 8 Xul	April 31, 820
9.19.15.0.0	1 Ahaw 3 Tzek	April 4, 825
10.0.0.0.0	7 Ahaw 18 Sip	March 9, 830
10.0.5.0.0	13 Ahaw 13 Woh	February 11, 835
10.0.10.0.0	6 Ahaw 8 Pop	January 16, 840
10.0.15.0.0	12 Ahaw 8 Kumk'u	December 20, 844
10.1.0.0.0	5 Ahaw 3 K'ayab	November 24, 849
10.1.5.0.0	11 Ahaw 18 Muwan	October 29, 854
10.1.10.0.0	4 Ahaw 13 K'ank'in	October 3, 859
10.1.15.0.0	10 Ahaw 8 Mak	September 6, 864
10.2.0.0.0	3 Ahaw 3 Yax	August 11, 869
10.2.5.0.0	9 Ahaw 18 Yax	July 16, 874
10.2.10.0.0	2 Ahaw 13 Ch'en	June 20, 879
10.2.15.0.0	8 Ahaw 8 Mol	May 24, 884
10.3.0.0.0	1 Ahaw 3 Yaxk'in	April 28, 889
10.3.5.0.0	7 Ahaw 18 Tzek	April 2, 894
10.3.10.0.0	13 Ahaw 13 Sotz'	March 7, 899
10.3.15.0.0	6 Ahaw 8 Sip	February 9, 904
10.4.0.0.0	12 Ahaw 3 Woh	January 13, 909

APPENDIX 5 The Rulers of Copan's Altar Q

RULER'S NAME	COSMOLOGICAL ORIENTATION	BAKTUN PERIOD
1. K'inich Yax K'uk' Mo'	W	9.0.0.0.0
2. K'inich(?) (Ruler 2)	W	9.0.0.0.0
3. ?	N	
4. K'al Tuun Hix	N	
5. ?	N	
6. Muyal Jol	N	
7. Bahlam Nehn	E	9.4.10.0.0
8. Wi' Ohl K'inich	E	9.5.0.0.0
		9.5.10.0.0
9. Sak?	E	
10. ? Bahlam	E	9.6.0.0.0
		9.6.10.0.0
11. K'ahk' Uti' Chan	S	9.9.0.0.0
		9.9.10.0.0
12. K'ahk' Uti' Witz' K'awiil	S	9.11.0.0.0
		9.11.15.0.0
		9.12.0.0.0
		9.12.10.0.0
		9.13.0.0.0
13. Waxaklajuun Ubaaj K'awiil	S	9.13.10.0.0
		9.14.0.0.0
		9.14.0.0
		9.14.10.0.0
		9.14.15.0.0
		9.15.0.0.0
		9.15.5.0.0
14. K'ahk' Yoplaj Chan K'awiil	S	
15. K'ahk' Yipyaj Chan K'awiil	W	9.16.5.0.0
		9.16.10.0.0
16. Yax Pasaj Chan Yopaat	W	9.17.0.0.0
		9.18.0.0.0
		9.19.0.0.0

ENDNOTES

1. The Itzá Prophecy

1. Quotation from Ralph L. Roys, *The Book of Chilam Balam of Chumayel*, 1967, reprinted (Washington, D.C.: Carnegie Institution, 1932), pp. 62, 169.
2. Juan de Villaguttiere Soto-Mayor, *History of the Conquest of the Province of the Itza*, trans. R.Wood (Culver City, CA: Labryinthos, 1983), p. 92.
3. Grant D. Jones, *The Conquest of the Last Maya Kingdom*, (Stanford, CA: Stanford University Press, 1998), p. 181.
4. Fray Andres Avendaño y Loyola, *Relation of Two Trips to Peten, Made for the Conversion of the Heathen Itzaex and Cehache*, trans. C. P. Bowditch and G. Rivera (Culver City, CA: Labryinthos Press, 1987), pp. 30–31.
5. Ibid., p. 39.
6. Roys, 1967, p. 136.
7. Ibid.
8. Ibid., p. 137.
9. See description of reposing *chilan* in Roys, 1967, p. 182.

2. Mesoamerican Times

1. From Bernal Díaz del Castillo, *The Discovery and Conquest of Mexico*, 2003, reprinted, trans. A. P. Maudslay (Boston: Da Capo Press, 1956), pp. 190–91.
2. See M. Rodríquez Martínez, et al., 2006, "The Earliest Writing in the New World," *Science* 313(5793), pp. 1610–1614.
3. See David Stuart, "The 'Arrival of Strangers': Teotihuacan and Tollan in Classic

Maya History," in *Mesoamerica's Classic Heritage: From Teotihuacan to the Aztecs*, eds. D. Carrasco, L. Jones, and S. Sessions (Boulder, CO: University Press of Colorado, 2000), pp. 465–514 and Charles C. Mann, *1491: New Revelations of the Americas Before Columbus* (New York: Vintage, 2006), pp. 273–76.

4. A. R. Williams, "Pyramid of Death," *National Geographic Magazine* 120:4 October 2006, pp. 144–153.

5. Fray Bernardino de Sahagún, *The Florentine Codex*, Book 7, trans. Arthur J. O. Anderson and C. Dibble (Santa Fe, NM: School of American Research and University of Utah Press, 1950–82), pp. 25, 27.

6. Miguel Leon-Portilla, *Aztec Thought and Culture* (Norman, OK: University of Oklahoma Press, 1974), pp. 14–15.

3. The Essence of Space

1. Ruth Bunzel, *Chichicastenango: A Guatemalan Village*, 2nd ed. (Seattle: University of Washington Press, 1959), p. 264.

2. Ruth Carlsen and Francis Eachus, "Kekchi Spirit World," in *Cognitive Studies in Southern Mesoamerica*, eds. Helen L. Neuenswander and Dean E. Arnold (Dallas, TX: Summer Institute of Linguistics Museum of Anthropology, 1977), pp. 38–65.

3. D. Boremanse, "Sewing Machines and Q'echi' Maya Worldview," *Anthropology Today* 16:1 (2000), pp. 11–19.

4. Edward B. Tylor, *Primitive Culture*, 1958, reprinted (New York: Harper and Row, 1871).

5. Carlos Lenkersdorf, *Cosmovisión Maya* (Mexico City: Centro de Estudios Antropológicos, Científicos, Artísticos Tradicionales y Lingüísticos Ce Acatl., 1999), p. 39.

6. A. R. Sandstrom, *Corn Is Our Blood: Culture and Ethnic Identity in a Contemporary Aztec Indian Village* (Norman, OK: University of Oklahoma Press, 1991), p. 241.

7. J. Richard Andrews and Ross Hassig (trans.), *Treatise on the Heathen Superstitions that Today Live among the Indians Native to this New Spain, 1629*, by Hernándo Ruiz de Alarcón (Norman, OK: University of Oklahoma Press, 1984), p. 48.

8. Eva Hunt, *The Transformation of the Hummingbird: Cultural Roots of a Zinacantecan Mythical Poem* (Ithaca, OK: Cornell University Press, 1977), p. 55.

9. Victor Parera and Robert Bruce, *The Last Lords of Palenque* (Berkeley, CA: University of California Press, 1982), p. 31.

10. Arild Hvidfeldt, *Teotl and Ixiptlatli: Some Central Conceptions in Ancient Mexican Religion, with a General Introduction on Cult and Myth*, (Copenhagen: Munksgaard, 1958).

11. *See* Evon Z. Vogt, *Tortillas for the Gods: a Symbolic Analysis of Zinacanteco Rituals* (Cambridge: Harvard University Press, 1976).

12. Robert Carlsen and Martin Prechtel, "The Flowering of the Dead: An Interpretation of Highland Maya Culture," *Man* 26 (1991), pp. 23–42.

13. J. L. Mondloch, "K'ex: Quiche Naming," *Journal of Mayan Linguistics* 1:2 (1980), 9–25.

14. Carlsen and Prechtel, 1991.

15. H. Wilbur Aulie and Evelyn Aulie, *Diccionario Ch'ol* (Mexico City: Insitutio Lingüístico del Verano, 1978).

16. Alfredo Lopez Austin, *The Human Body and Ideology: Concepts of the Ancient Nahuas*, 2 vols. (Salt Lake City, UT: University of Utah Press, 1988).

17. Vogt, 1976; William F. Hanks, *Referential Practice. Language and Lived Space among the Maya* (Chicago: University of Chicago Press, 1990), p. 300.

18. Hanks, 1990, pp. 352–80.

19. Alfred Kroeber, *Anthropology* (New York: Harcourt and Brace, 1923), p. 252.

20. Andrews and Hassig, 1984, p. 72.

21. Michael D. Coe, "Community Structure," *Southwestern Journal of Anthropology*, 21:2 (1965), pp. 97–114.

22. Barbara Tedlock, "The Road of Light: Theory and Practice of Mayan Skywatching," in *The Sky in Mayan Literature*, ed. A. F. Aveni (New York: Oxford University Press, 1992), pp. 18–42.

23. W. F. Morris and J. J. Fox, *Living Maya* (New York: Henry N. Abrams, 2000), p. 73.

24. Diego García de Palacio, *Letter to the King of Spain: Being a Description of the Ancient Provinces of Guazacapan, Izalco, Cuscatlan, and Chiquimula, in the Audiencia of Guatemala, with an Account of the Languages, Customs, and Religion of their Aboriginal Inhabitants, and a Description of the Ruins of Copan* (Culver City, CA: Labyrinthos, 1985), pp. 37–39.

25. Alan R. Sandstrom, *Corn Is Our Blood: Culture and Ethnic Identity in a Contemporary Aztec Indian Village* (Norman, OK: University of Oklahoma Press, 1991), p. 239.

26. Robert Carlsen, *The War for the Heart and Soul of a Highland Maya Town* (Austin, OK: University of Texas Press, 1997), p. 52.

27. Ibid., p. 53; Allen J. Christenson, *Art and Society in a Highland Maya Community: The Altarpiece of Santiago Atitlán* (Austin, TX: University of Texas Press, 2001).

28. Sandstrom, 1991, p. 239.

29. R. Lok, "The House as a Microcosm. Some Cosmic Representations in a Mexican Village," in *The Leiden Tradition in Structural Anthropology*, eds. R. de Ridder and J. Karremans (Leiden, The Netherlands: E. J. Brill, 1987), pp. 211–223.

30. Stephen Houston and David Stuart, "Peopling the Maya Court," in T. Inomata and S. Houston, eds., *Royal Courts of the Ancient Maya, Vol 1: Theory, Comparison, and Synthesis* (Boulder, CO: Westview Press, 2001), pp. 54–83.

4. Finding Order

1. Anthony Aveni, *Empires of Time* (Boulder, CO: University Press of Colorado, 2002), p. 27.

5. Ideas of the Day

1. Sahagún, *The Florentine Codex*, Book 4, p. 23.

2. Fray Diego de Hevia y Valdes, *Relación autentica de las idolatrías, supersticiones, vanas observaciones de los indios de obispado de Oaxaca* (Mexico City, 1656), p. 187.

3. Fray Diego Durán, *Book of the Gods and Rites and the Ancient Calendar* (Norman, OK: University of Oklahoma Press, 1971), p. 396.

4. Ibid., p. 397.

5. Ibid.

6. Ibid., p. 386.

7. Sahagún, *The Florentine Codex*, Book 10, p. 31.

8. Sandstrom, 1991, p. 264.

9. Bunzel, 1959, p. 277.

10. Vicente Pineda, *Historia de las sublevaciones indígenas habidas en el estado de Chiapas*, 1986, reprinted (Mexico City: Instituto Nacional Indigenista, 1888).

11. O. LaFarge and D. Beyers, *The Year Bearer's People* (New Orleans, LA: Tulane University, 1931), p. 176.

12. From Blom's letter of presentation in Ibid., p. xiii.

13. J. E. S. Thompson, *Maya History and Religion* (Norman, OK: University of Oklahoma Press, 1970), p. xv.

14. Maud Oakes, *The Two Crosses of Todos Santos: Survivals of Maya Religious Ritual* (New York: Pantheon Books, 1951).

15. Charles P. Bowditch, *The Numeration, Calendar Systems and Astronomical Knowledge of the Mayas* (Cambridge: The University Press, 1910), p. 267.

16. J. E. S. Thompson, *Maya Hieroglyphic Writing: An Introduction* (Washington, D.C.: Carnegie Institution of Washington, 1950), p. 98.

17. Leonhard Schultze Jena's words, quoted in Peter T. Furst, "Human Biology and the Origin of the 260-Day Sacred Almanac: The Contributions of Leonhard Schultze Jena (1872–1955)," in G. Gossen, *Symbol and Meaning Beyond the Classed Community: Essays in Mesoamerican Ideas* (Albany, NY: Institute for Mesoamerican Studies, SUNY Albany, 1986), pp. 69–76.

18. Gary Gossen, "A Chamula Calendar Board from Chiapas, Mexico," in *Meso-American Archaeology: New Approaches*, ed. Norman Hammond (Austin, TX: University of Texas Press, 1974), pp. 217–254.

19. Tedlock, 1992, p. 26.

6. Long Counting

1. Michael Coe, "Royal Fifth: Earliest Notices of Maya Writing," *Research Reports on Ancient Maya Writing*, 28 (Washington, D.C.: Center for Maya Research, 1989).

2. J. T. Goodman, *The Archaic Maya Inscriptions* (London, England: R.H. Porter and Dulau, 1897).

3. J. Martínez Hernández, *Paralelismo entre los calendarios maya y azteca. Su correlación con el calendario juliano* (Merida: Compañia Tipográfica Yucateca, 1926); J. E. S. Thompson, "Maya Chronology: The Correlation Question," *Carnegie Institution of Washington*, Pub. 456, Contrib. 14 (Washington, D.C.: Carnegie Institution of Washington, 1935).

4. John Edgar Teeple, "Maya Astronomy," *Carnegie Institution of Washington Contributions to American Archaeology*, vol. 1, no. 2 (Washington, D.C.: Carnegie Institution of Washington, 1930), pp. 29–115.

5. Floyd G. Lounsbury, "A Derivation of the Mayan-to-Julian Calendar Correlation from the Dresden Codex Venus Chronology," in *The Sky in Mayan Literature*, ed. A. F. Aveni (New York: Oxford University Press, 1992), pp. 184–206.

6. Tedlock, 1982, p. 268.

7. Anthony Aveni, *Skywatchers* (Austin, TX: University of Texas Press, 2001), p. 171.

7. Beginnings and Endings of the World

1. Roys, 1967, pp. 101–2.
2. Miguel Leon-Portilla, *The Broken Spears: The Aztec Account of the Conquest of Mexico* (Boston: Beacon Books, 1992), p. 149.
3. Allen J. Christenson, *Popol Vuh: The Sacred Book of the Maya* (Winchester: O Books, 2003), p. 65.
4. Ibid., pp. 67–68.
5. Ibid., p. 71.
6. Ibid., p. 82.
7. Ibid., p. 92.
8. B. Tedlock, 1992, p. 27.

8. The Deepest Time

1. Roys, 1967, p. 98.
2. Matthew Looper, *Lightning Warrior: Maya Art and Kingship at Quirigua* (Austin, TX: University of Texas Press, 2003) also offers an excellent analysis of the iconography on the monuments of Quiriguá.

9. Kings of Time

1. From D. López de Cogolludo, 1868(1688), *Historia de Yucatan*, Book 4, Chapter 8. Quoted by R. L. Roys, 1967, p. 131.
2. See Jared Diamond, *Collapse: How Societies Choose to Fail or Succeed* (London, England: Penguin Books, 2005), p. 439.
3. Rafael Girard, *Los Mayas: su civilización, su historia, sus vinculaciones continentales* (Mexico City: Libro Mex, 1966), p. 32.
4. Aulie and Aulie, 1978.
5. Roys, 1967, p. 184.
6. D. Puleston, *Maya Arcaheology and Ethnohistory* (Austin, TX: University of Texas Press).

10. Seeing Stars

1. Stephen Jay Gould, *Questioning the Millennium: A Rationalist's Guide to a Precisely Arbitrary Countdown* (New York: Harmony Books, 1997), pp. 157–58.
2. Sylvanus G. Morley, "The Foremost Intellectual Achievement of Ancient America, the Hieroglyphic Inscriptions on the Monuments in the Ruined Cities of Mexico, Guatemala, and Honduras Are Yielding the Secrets of the Maya Civilization," *National Geographic Magazine*, February, 1922: 109–130, p. 125.
3. Aveni, 2001, p. 13.
4. Ibid., pp. 184–96.

5. Sylvanus G. Morley, *The Ancient Maya* (Stanford: Stanford University Press, 1946), p. 70.

6. F. G. Lounsbury, "Astronomical Knowledge and Its Uses at Bonampak, Mexico," in *Archaeoastronomy in the New World*, ed. A. Aveni, (Cambridge: Cambridge University Press, 1982), pp. 143–168.

7. Tedlock, 1992, p. 28.

8. David L. Freidel, Linda Schele and Joy Parker, *Maya Cosmos: Two Thousand Years on the Shaman's Path* (New York: William Morrow, 1993), p. 79.

9. Ibid., pp. 65–68.

10. Ibid., p. 87.

11. Daniel Pinchbeck, *2012: The Return of Quetzalcoatl* (New York: Penguin, 2006).

12. Carl Johan Calleman, *The Mayan Calendar and the Transformation of Consciousness* (Rochester, VT: Bear and Co., 2004).; Lawrence E. Joseph, *Apocalypse 2012: A Scientific Investigation Into Civilization's End* (New York: Morgan Road Books, 2007).

13. John Major Jenkins, *Maya Cosmogenesis 2012* (Rochester, VT: Bear and Co., 1998), p. 332.

14. John Major Jenkins, "The Origins of the 2012 Revelation," in *The Mystery of 2012: Predictions, Prophecies and Possibilities* (Boulder, CO: Sounds True, 2007), p.44.

BIBLIOGRAPHY

Aldana, G. 2005. "Agency and the 'Star War' Glyph: A Historical Reassessment of Classic Maya Astrology and Warfare," *Ancient Mesoamerica* 16:305–320.

Ambrose, S. E. 1996. *Undaunted Courage: Meriwether Lewis, Thomas Jefferson, and the Opening of the American West*. New York: Simon and Schuster.

Andrews, J. R., and R. Hassig (trans.). 1984. *Treatise on the Heathen Superstitions that Today Live among the Indians Native to this New Spain, 1629*, by Hernándo Ruiz de Alarcón. Norman, OK: University of Oklahoma Press.

Aulie, H. W., and E. Aulie. 1978. *Diccionario Ch'ol*. Mexico City: Insitutio Lingüistico del Verano (Summer Institute of Linguistics).

Avendaño y Loyola, Fray A. 1987. *Relation of Two Trips to Peten, Made for the Conversion of the Heathen Itzaex and Cehaches*. Translated by C. P. Bowditch and G. Rivera. Culver City, CA: Labryinthos Press.

Aveni, A. 2001. *Skywatchers*. Austin, TX: University of Texas Press.

—————. 2002. *Empires of Time*. Boulder, CO: University Press of Colorado.

Aveni, A., and L. D. Hotaling. 1994. "Monumental Inscriptions and the Observational Basis of Maya Planetary Astronomy," *Archaeoastronomy* 19:S21–S54.

Berrin, K. and E. Pasztory. 1993. *Teotihuacan: Art from the City of the Gods*. London, England: Thames and Hudson.

Boone, E. H. 2007. *Cycles of Time and Meaning in the Central Mexican Books of Fate*. Austin, TX: University of Texas Press.

Boremanse, D. 2000. "Sewing Machines and Q'echi' Maya Worldview," *Anthropology Today* 16 (1):11–19.

Bowditch, C.P. 1910. *The Numeration, Calendar Systems and Astronomical Knowledge of the Mayas*. Cambridge: The University Press.

Brinton, D. G. 1895. *A Primer of Mayan Hieroglyphics*. Philadelphia, PA: University of Pennsylvania Press.

Bunzel, R. 1959. *Chichicastenango: A Guatemalan Village*, 2nd ed. Seattle, WA: University of Washington Press.

Calleman, C. J. 2004. *The Mayan Calendar and the Transformation of Consciousness*. Rochester, VT: Bear and Co.

Carlsen, R. 1997. *The War for the Heart and Soul of a Highland Maya Town*. Austin, TX: University of Texas Press.

Carlsen R., and F. Eachus. 1977. "Kekchi Spirit World" in *Cognitive Studies in Southern Mesoamerica*, Helen L. Neuenswander and Dean E. Arnold, eds., pp. 38–65. Dallas, TX: Summer Institute of Linguistics Museum of Anthropology.

Carlsen, R., and M. Prechtel. 1991. "The Flowering of the Dead: An Interpretation of Highland Maya Culture," *Man* 26:23–42.

Caso, A. 1967. *Los calendarios prehispánicos*. Mexico City: UNAM.

Chinchilla Mazariegos, O. 2006. A Reading of the "Earth-Star" Verb in Ancient Maya Writing. *Research Reports on Ancient Maya Writing* 56. Barnardsville, NC: Center for Maya Research.

Christenson, A. J. 2001. *Art and Society in a Highland Maya Community: The Altarpiece of Santiago Atitlán*. Austin, TX: University of Texas Press.

_____. 2003. *Popol Vuh, The Sacred Book of the Maya*. Hampshire, UK: O Books.

Coe, M. D. 1965. "Community Structure," *Southwestern Journal of Anthropology*, 21(2):97–114.

_____. 1989. "Royal Fifth: Earliest Notices of Maya Writing," *Research Reports on Ancient Maya Writing*, 28. Washington, D.C.: Center for Maya Research.

_____. 1992. *Breaking the Maya Code*. New York: Thames and Hudson.

_____. 2005. *The Maya*. London, England: Thames and Hudson.

Coe, M.D., and M. Van Stone. 2001. *Reading the Maya Glyphs*. London, England: Thames and Hudson.

Coggins, C. C. 1980. "The Shape of Time: Some Political Implications of a Four-Part Figure," *American Antiquity* 45(4):727–739.

Colby, B. N., and L. M. Colby. 1981. *The Daykeeper: The Life and Discourse of an Ixil Diviner*. Cambridge: Harvard University Press.

Cutler, et al., 1987. "Lunar Influences in the Reproductive Cycle in Women," *Human Biology* 59 (6):959–972.

DeBoer, W. R. 2005. "Colors for a North American Past," *World Archaeology* 37(1):66–91.

Diamond, J. 2005. *Collapse: How Societies Choose or Fail to Succeed*. London, England: Penguin Books.

Díaz del Castillo, B. 2003. *The Discovery and Conquest of Mexico*. Boston: Da Capo Press.

Diehl, R. A. 2004. *The Olmecs: America's First Civilization*. London, England: Thames and Hudson.

Durán, Fray D. 1964. *The Aztecs: the History of the Indies of New Spain*. London, England: Cassell.

_____. 1971. *Book of the Gods and Rites and the Ancient Calendar*. Norman, OK: University of Oklahoma Press.

Edmonson, M.S. 1979. "Some Postclassic Questions about the Classic Maya," in M. Robertson and D. Jeffers, *Tercera Mesa Redonda de Palenque*, pp. 9–18. Palenque, Mexico: Pre-Columbian Art Research Center.

_____. 1982. *The Ancient Future of the Itza: The Book of Chilam Balam of Tizimin*. Austin, TX: University of Texas Press.

_____. 1986. *Heaven Born Merida and its Destiny: the Book of Chilam Balam of Chumayel*. Austin, TX: University of Texas Press.

Elson, C., and M. E. Smith. 2001. "Archaeological Deposits from the Aztec New Fire Ceremony," *Ancient Mesoamerica* 12(2):157–174.

Evans, R.T. 2004. *Romancing the Maya: Mexican Antiquity in the American Imagination, 1820–1915*. Austin, TX: University of Texas Press.

Fash, W. 2001. *Scribes, Warriors and Kings: The City of Copan and the Ancient Maya*, 2nd ed. London, England: Thames and Hudson.

Freidel, D., L. Schele and J. Parker. 1993. *Maya Cosmos: Two Thousand Years on the Shaman's Path*. New York: William Morrow.

Furst, P. T. 1986. "Human Biology and the Origin of the 260-Day Sacred Almanac: The Contributions of Leonhard Schultze Jena (1872–1955)," in G. Gossen, *Symbol and Meaning Beyond the Closed Community: Essays in Mesoamerican Ideas*, pp. 69–76. Albany, NY: Institute for Mesoamerican Studies, SUNY Albany.

Diego García de Palacio. 1985. *Letter to the King of Spain: Being a Description of the Ancient Provinces of Guazacapan, Izalco, Cuscatlan, and Chiquimula, in the Audiencia of Guatemala, with an Account of the Languages, Customs, and Religion of their Aboriginal Inhabitants, and a Description of the Ruins of Copan*. Culver City, CA: Labyrinthos.

Geertz, A. W. "Contemporary Problems in the Study of Native North American Religions with Special Reference to the Hopis," *American Indian Quarterly* 20(3):393–414.

Girard, R. 1966. *Los Mayas: su civilización, su historia, sus vinculaciones continentales*. Mexico City: Libro Mex.

Girard, R., and B. Preble. 1995. *People of the Chan*. Chino Valley, AZ: Continuum Foundation.

Goodman, J. T. 1897. *The Archaic Maya Inscriptions*. London, England: R.H. Porter and Dulau.

_____. 1905. "Maya Dates," *American Anthropologist* 7(4):642–647.

Gossen, G. 1974. "A Chamula Calendar Board from Chiapas, Mexico," in *Meso-American Archaeology: New Approaches*, Norman Hammond, ed., pp. 217–254. Austin, TX: University of Texas Press.

Gould, S. J. 1988. *Time's Arrow, Time's Cycle. Myth and Metaphor in the Discovery of Geologic Time*. Cambridge: Harvard University Press.

_____. 1997. *Questioning the Millennium: A Rationalist's Guide to a Precisely Arbitrary Countdown*. New York: Harmony Books.

Graham, I., and E. von Euw. 1997. *Corpus of Maya Hieroglyphic Inscriptions, Volume 8, Number 1: Coba*. Cambridge, MA: Peabody Museum, Harvard University.

Gronemeyer, S. 2006. *The Maya Site of Tortuguero, Tabasco, Mexico: Its History and Inscriptions*. Acta Mesoamericana, vol. 17. Markt Schwaben, Germany: Verlag Anton Saurwein.

Gubler, R. 1997. "The Importance of the Number Four as an Ordering Principle in the World View of the Ancient Maya," *Latin American Indian Literatures Journal* 13(1):23–57.

Hanks, W. F. 1990. *Referential Practice. Language and Lived Space among the Maya*. Chicago: University of Chicago Press.

Harris III, C. H., and L. R. Sadler. 2003. *The Archaeologist Was a Spy: Sylvanus G.*

Morley and the Office of Naval Intelligence. Albuquerque, NM: University of New Mexico Press.

Hevia y Valdes, Fray D. de. 1666. *Relación autentica de las idolatrias, supersticiones, vanas observaciones de los indios de obispado de Oaxaca*. Mexico City.

Hirth, K. 2000. *Archaeological Research at Xochicalco*. Salt Lake City, UT: University of Utah Press.

Hopkins, N., A. Cruz Guzman and J.K. Josserand. 2008. "A Chol (Mayan) Vocabulary from 1789." *International Journal of American Linguistics* 74(1):83–114.

Horcasitas, F, and D. Heyden. 1971. Introduction to Durán, 1971.

Houston, S. D. 2009. "What Won't Happen in 2012;" http://decipherment.wordpress.com/2008/12/20/what-will-not-happen-in-2012.

Houston, S. D., and T. Inomata. 2009. *The Classic Maya*. Cambridge: Cambridge University Press.

Houston, S. D., and D. Stuart. 2001. "Peopling the Maya Court," in *Royal Courts of the Ancient Maya, Vol 1: Theory, Comparison, and Synthesis*, T. Inomata and S. Houston, eds., pp. 54–83. Boulder: Westview Press.

Hunt, E. 1977. *The Transformation of the Hummingbird: Cultural Roots of a Zinacantecan Mythical Poem*. Ithaca, NY: Cornell University Press.

Hvidfeldt, A. 1958. *Teotl and Ixiptlatli: Some Central Conceptions in Ancient Mexican Religion, with a General Introduction on Cult and Myth*, Copenhagen: Munksgaard.

Jenkins, J. M. 1998. *Maya Cosmogenesis 2012*. Rochester, VT: Bear and Co.

_____. 2007. "The Origins of the 2012 Revelation," in *The Mystery of 2012: Predictions, Prophecies and Possibilities*. Boulder, CO: Sounds True.

Jones, C., and L. Satterthwaite. 1980. *The Monuments and Inscriptions of Tikal: The Carved Monuments*. Tikal Reports, no. 33, Part A. Philadelphia, PA: The University Museum, University of Pennsylvania.

Jones, G. D. 1989. *Maya Resistance to Spanish Rule*. Albuquerque, NM: University of New Mexico Press.

_____. 1998. *The Conquest of the Last Maya Kingdom*. Stanford, CA: Stanford University Press.

Joseph, L. E. 2007. *Apocalypse 2012: A Scientific Investigation Into Civilization's End*. New York: Morgan Road Books.

Kelley, D. H. 1983. "The Maya Calendar Correlation Problem," in *Civilization in the Ancient Americas: Essays in Honor of Gordon R. Willey*, R. Levanthal and A. Kolata, eds., pp. 157–208. Albuquerque, NM: University of New Mexico Press.

Kirchoff, P. 1943. "Mesoamérica: Sus Límites Geográficos, Composición Étnica y Caracteres Culturales," *Acta Americana* 1(1):92–107.

Knab, T. J. 1995. *A War of the Witches: A Journey into the Underworld of the Contemporary Aztecs*. Boulder, CO: Westview Press.

Kroeber, A. 1923. *Anthropology*. New York: Harcourt and Brace.

La Farge, O. 1934. "Post-Columbian Dates and the Mayan Correlation Problem," *Maya Research* 1:109–124.

_____. 1947. *Santa Eulalia: The Religion of a Cuchumatán Indian Town*. Chicago: University of Chicago Press.

La Farge, O., and D. Beyers. 1931. *The Year Bearer's People*. New Orleans, LA: Tulane University.

Landa, Fray D. de. 1941. *Relacion de las Cosas de Yucatan*. Translated by A. M. Tozzer, Cambridge: Peabody Museum, Harvard University.

Lenkersdorf, C. 1999. *Cosmovisión Maya*. Mexico City: Centro de Estudios Antropológicos, Científicos, Artísticos Tradicionales y Lingüísticos Ce Acatl.

Leon-Portilla, M. 1974. *Aztec Thought and Culture*. Norman, OK: University of Oklahoma Press.

_____. 1992. *The Broken Spears: The Aztec Account of the Conquest of Mexico*. Boston: Beacon Books.

Lipp, F. J. 1991. *The Mixe of Oaxaca: Religion, Ritual and Healing*. Austin, TX: University of Texas Press.

_____. 1997. *Cycles of the Sun, Mysteries of the Moon*. Austin, TX: University of Texas Press.

Lok, R. 1987. "The House as a Microcosm. Some Cosmic Representations in a Mexican Village," in *The Leiden Tradition in Structural Anthropology*, R. de Ridder and J. Karremans, eds., pp. 211–223. Leiden, The Netherlands: E. J. Brill.

Looper, M. 2003. *Lightning Warrior: Maya Art and Kingship at Quirigua*. Austin, TX: University of Texas Press.

López de Cogolludo, D. 1868 (1688). *Historia de Yucatan*. Madrid: Juan Garcia Infanzen.

López Austin, A. 1988. *The Human Body and Ideology: Concepts of the Ancient Nahuas*. 2 vols. Salt Lake City, UT: University of Utah Press.

López Luján, L. 2008. " 'El adiós y triste queja del Gran Calendario Azteca,' El incesante peregrinar de la Piedra del Sol," *Arqueología Mexicana*, vol. XV(91): 78–83.

Lounsbury, F. G. 1978. "Maya Numeration, Computation, and Calendrical Astronomy," in *Dictionary of Scientific Biography*, C.C. Gillespie, ed., vol. 15, sup. 1, pp. 759–818. New York: Charles Scribner and Sons.

_____. 1982. "Astronomical Knowledge and Its Uses at Bonampak, Mexico," in *Archaeoastronomy in the New World*, A. Aveni, ed., pp. 143–168. Cambridge: Cambridge University Press.

_____. 1992. "A Derivation of the Mayan-to-Julian Calendar Correlation from the Dresden Codex Venus Chronology," in *The Sky in Mayan Literature*, A. F. Aveni, ed., pp. 184–206. New York: Oxford University Press.

Lowe, G. W. 1962. "Algunos resultados de la temporada 1961 en Chiapa de Corzo, Chiapas," *Estudios de Cultura Maya* 2:185–196. Mexico City: UNAM.

Mann. C. C. 2006. *1491: New Revelations of the Americas Before Columbus*. New York: Vintage.

Marcus, J., and K. Flannery. 1996. *Zapotec Civilization: How Urban Society Evolved in Mexico's Oaxaca Valley*. London, England: Thames and Hudson.

Martin, S., and N. Grube. 2000. *Chronicles of Maya Kings and Queens: Deciphering the Dynasties of the Ancient Maya*. London, England: Thames and Hudson.

Martínez Hernández, J. 1926. *Paralelismo entre los calendarios maya y azteca. Su correlación con el calendario juliano*. Merida: Compañia Tipográfica Yucateca.

Matos, E., and F. Solis. 2004. *El Calendario Azteca y Otros Monumentos Solares*. Mexico City: CONACULTA-Instituto Nacional de Antropologia e Historia.

Menaker W., and A. Manaker. 1959. "Lunar Periodicity in Human Reproduction: A Likely Unit of Biological Time," *American Journal of Obstetrics and Gynecology* 77(4): 905–914.

Metz, Brent, et. al. 2009. *The Ch'orti' Maya Area Past and Present*. Gainesville, FL: University Press of Florida.

Mondloch, J. L. 1980. "K'ex: Quiche Naming," *Journal of Mayan Linguistics* 1(2):9–25.

Montgomery, J. 2002. *How to Read Maya Hieroglyphs*. New York: Hippocrene Books.

Morley, S. G. 1915. *An Introduction to the Study of Maya Hieroglyphs*. Bureau of American Ethnology, Bulletin 57. Washington, D.C.: Government Printing Office.

_____. 1920. *The Inscriptions at Copan*. Washington, D.C.: Carnegie Institution of Washington.

_____. 1922. "The Foremost Intellectual Achievement of Ancient America, the Hieroglyphic Inscriptions on the Monuments in the Ruined Cities of Mexico, Guatemala, and Honduras are Yielding the Secrets of the Maya Civilization," *National Geographic Magazine*, February, pp. 109–130.

_____. 1946. *The Ancient Maya*, 1st ed. Stanford, CA: Stanford University Press.

Morris, W. F., and J. J. Fox. 2000. *Living Maya*. New York: Henry N. Abrams.

Nahm, W. 1994. "Maya Warfare and the Venus Year," *Mexicon* 16(1):6–10.

Nuttall, Z. 1927. "Origin of the Maya Calendar," *Science* 65(1678):12–14.

Oakes, M. 1951. *The Two Crosses of Todos Santos: Survivals of Maya Religious Ritual*. New York: Pantheon Books.

Pagden, A. (trans.) 1986. *Hernan Cortes: Letters from Mexico*. New Haven, CT: Yale University Press.

Parera, V., and R. Bruce. 1982. *The Last Lords of Palenque*. Berkeley, CA: University of California Press.

Pinchbeck, D. 2006. *2012: The Return of Quetzalcoatl*. New York: Penguin.

Pineda, V. 1986 (1888). *Historia de las sublevaciones indígenas habidas en el estado de Chiapas*. Mexico City: Instituto Nacional Indigenista.

Pool, C. 2007. *Olmec Archaeology and Early Mesoamerica*. Cambridge: Cambridge University Press.

Proskouriakoff, T. 1960. "Historical Implications of a Pattern of Dates at Piedras Negras, Guatemala," *American Antiquity* 25(4):454–475.

Puleston, D. 1977. In *Maya Archaeology and Ethnohistory*, N. Hammond, ed., pp. 63–74. Austin, TX: University of Texas Press.

Restall, M. 1998. *Maya Conquistador*. Boston: Beacon Press.

Rice, P. M. 2004. *Maya Political Science: Time, Astronomy, and the Cosmos*. Austin, TX: University of Texas Press.

_____. 2008. *Maya Calendar Origins: Monuments, Mythistory and the Mateiralization of Time*. Austin, TX: University of Texas Press.

Rodríguez Martínez, M. de, P. Ortíz Ceballos, M. D. Coe, R. A. Deihl, S. D. Houston, K. A. Taube and A. Delgado Calderón. 2006. "The Earliest Writing in the New World," *Science* 313(5793):1610–1614.

Roys, R. L. 1967 (1932). *The Chilam Balam of Chumayel*. Norman, OK: University of Oklahoma Press.

Ruiz de Alarcón, Hernándo. 1984. *Treatise on the Heathen Superstitions that Today Live among the Indians Native to this New Spain, 1629*. Translated by J. R. Andrews and R. Hassig. Norman, OK: University of Oklahoma Press.

Sahagún, Fray B. de. 1950–82. *The Florentine Codex*. Translated by A. J. O. Anderson and C. Dibble. Santa Fe, NM: School of American Research and University of Utah Press.

Sandstrom, A. R. 1991. *Corn Is Our Blood: Culture and Ethnic Identity in a Contemporary Aztec Indian Village*. Norman, OK: University of Oklahoma Press.

Saturno, W. 2003. "Sistine Chapel of the Early Maya," *National Geographic Magazine* 204(6):72–77.

_____. 2006. "Dawn of Maya Gods and Kings," *National Geographic Magazine* 209(1):68–77.

Saturno, W., K. Taube, and D. Stuart. 2005. "The Murals of San Bartolo, El Petén, Guatemala, Part 1: The North Wall," *Ancient America*, no. 7. Barnardsville, NC: Boundary End Archaeological Research Center.

Shady, R. Haas, and J. Creamer, W. 2001. "Dating Caral, a Preceramic Site in the Supe Valley on the Central Coast of Peru," *Science* 292:723–726.

Sharer, R. J, and L. P. Traxler. 2006. *The Ancient Maya*, 6th ed. Stanford, CA: Stanford University Press.

Sitler, R.K. 2006. "The 2012 Phenomenon: New Age Appropriation of an Ancient Mayan Calendar," *Novo Religio: the Journal of Alternative and Emergent Religions* 3:24–38.

Smith, A. L. 1950. *Uaxactun, Guatemala. Excavations of 1931–1937.* Washington, D.C., Carnegie Institution of Washington.

Solomon, C. 2002. *Tatiana Proskouriakoff: Interpreting the Ancient Maya.* Norman, OK: University of Oklahoma Press.

Stephens, J. L. 1839. *Incidents of Travel in Central America, Chiapas and Yucatan.* New York: Harper and Bros.

_____. 1841. *Incidents of Travel in Yucatan.* New York: Harper and Bros.

Stirling, M. 1940. An Initial Series from Tres Zapotes, Veracruz. *Contributed Technical Papers, Mexican Archaeology Series* 1(1). Washington, D.C.: National Geographic Society.

Stuart, D. 1993. "Historical Inscriptions and the Maya Collapse," in *Late Lowland Maya Civilization in the Eighth Century A.D.,* ed. by J. A. Sabloff and J. S. Henderson, pp. 321–354. Washington, D.C.: Dumbarton Oaks.

_____. 2000. "The 'Arrival of Strangers': Teotihuacan and Tollan in Classic Maya History," in *Mesoamerica's Classic Heritage: From Teotihuacan to the Aztecs,* D. Carrasco, L. Jones, and S. Sessions, eds., pp. 465–514. Boulder, CO: University Press of Colorado.

_____. 2006. *The Inscriptions of Temple XIX at Palenque.* San Francisco, CA: Pre-Columbian Art Research Institute.

Stuart, G. E. 1992. "Mural Masterpieces of Ancient Cacaxtla," *National Geographic Magazine* 182(3):120–136.

_____. 1992. "Quest for Decipherment: A Historical and Bibliographical Survey of Maya Hieroglyphic Investigation," in *New Theories on the Ancient Maya*, E. Danien and R. Sharer, eds., pp. 1–64. Philadelphia, PA: The University Museum, University of Pennsylvania.

_____. 1997. "The Royal Crypts of Copan," *National Geographic Magazine* 192(6):68–73.

Stuart, D., and G. Stuart. 2008. *Palenque: Eternal City of the Maya.* New York: Thames and Hudson.

Taube, K. A. 1992. *The Major Gods of Ancient Yucatan.* Studies in Pre-Columbian Art and Archaeology, 32. Washington, D.C.: Dumbarton Oaks.

Tedlock, B. 1982. *Time and the Highland Maya.* Albuquerque, NM: University of New Mexico Press.

_____. 1992. "The Road of Light: Theory and Practice of Mayan Skywatching," in *The Sky in Mayan Literature*, A. F. Aveni, ed., pp. 18–42. New York: Oxford University Press.

Teeple, J. E. 1930. "Maya Astronomy." *Carnegie Institution of Washington Contributions to American Archaeology*, vol. 1, no. 2, pp. 29–115. Washington, D.C.: Carnegie Institution of Washington.

Thompson, J. E. S. 1935. "Maya Chronology: The Correlation Question," in American Archaeology no. 14, Publication 456, pp. 51–104. Washington, D.C.: Carnegie Institution of Washington.

_____. 1938. *Correlation of Maya and Christian Chronologies*. Washington, D.C.: Carnegie Institution of Washington.

_____. 1950. *Maya Hieroglyphic Writing: An Introduction*. Washington, D.C.: Carnegie Institution of Washington.

_____. 1970. *Maya History and Religion*. Norman, OK: University of Oklahoma Press.

Tylor, E. B. 1958 (1871). *Primitive Culture*. New York: Harper and Row.

Umburger, E. 1986. Events Commemorated by Date Plaques at the Templo Mayor: Further Thoughts on the Solar Metaphor. In *The Aztec Templo Mayor*, E. H. Boone, ed., pp. 411–450. Washington, D.C.: Dumbarton Oaks.

Velázquez, P. F. (trans.). 1975. *Codice Chimalpopoca, Anales de Cuauhtitlan y Leyenda de los Soles*. Mexico, D.F.: UNAM.

Villaguttiere Soto-Mayor, J. de. 1983. *History of the Conquest of the Province of the Itza*. Translated by R.Wood. Culver City, CA: Labryinthos.

Villela, K., and M. E. Miller. 2010. *The Aztec Calendar Stone*. Los Angeles, CA: The Getty Research Institute.

Vogt, E.Z. 1969. *Zinacantan: A Maya Community in the Highlands of Chiapas*. Cambridge: Harvard University Press.

_____. 1976. *Tortillas for the Gods: a Symbolic Analysis of Zinacanteco Rituals*. Cambridge: Harvard University Press.

Wauchope, R. 1962. *Lost Tribes and Sunken Continents*. Chicago: University of Chicago Press.

Webster, David. 1999. "The Archaeology of Copan, Honduras," *Journal of Archaeological Research* 7(1):1–53.

_____. 2002. *The Fall of the Ancient Maya*. London, England: Thames & Hudson.

Wheatley, P. 1971. *The Pivot of the Four Quarters*. Edinburgh, UK: Edinburgh University Press.

Williams, A. R. 2006. "Pyramid of Death," *National Geographic Magazine* 120(4): 144–153.

Yasugi, Y. 1995. *"Native Middle American Languages: An Areal-Typological Perspective,"* in Senri Ethnological Studies, no. 39. Osaka, Japan: National Museum of Ethnology.

ACKNOWLEDGMENTS

I owe my thanks and appreciation to many colleagues, students, friends, and family members who, knowingly or not, helped in the preparation and writing of this book. The varied ideas that found their way into these pages developed over the course of many years, fostered and inspired by many teachers and fellow researchers in the field of Mesoamerican studies. Among them are Will Andrews, Barbara Arroyo, Anthony Aveni, Karen Bassie-Sweet, Elizabeth Boone, Davíd Carrasco, Oswaldo Chinchilla, Michael Coe, Arthur Demarest, Hector Escobedo, Barbara Fash, William Fash, David Freidel, Ian Graham, Sven Gronemeyer, Nikolai Grube, Julia Guernsey, John Hoopes, Heather Hurst, Grant Jones, Alfonso Lacadena, Leonardo López Luján, Barbara MacLeod, Simon Martin, Peter Mathews, John Monaghan, Merle Greene Robertson, Alan Sandstrom, Bill Saturno, Robert Sharer, Joel Skidmore, Emily Umberger, David Webster, and Marc Zender. A special debt of gratitude goes to my very close friends and colleagues Stephen Houston and Karl Taube, who over the years have inspired so much of what I think I know about archaeology, the Maya, and Mesoamerica.

Despite being gone now for many years, two important people left lasting marks on my understanding of the worlds of the ancient and modern Maya. The late Evon Vogt, Professor Emeritus of Anthropology at Harvard University, was a true inspiration even before I could call him a dear friend and collaborator; I will always have fond memories of our weekly lunches at Harvard. And the great Linda Schele, my mentor from the early days, planted so many insights and hunches in my mind that still play themselves out, leading me always in new directions of research. We didn't always agree with one another, but in a remarkable way she is still teaching me.

I have long appreciated the help, influence, and patience of many of my students past and present, including Gerardo Aldana, Nick Carter, Caitlin Earley, James Fitzsimmons, Tom Garrison, Lucia Henderson, Sarah Jackson, Kimberley Jones, Danny Law, Allan Maca, Edwin Román, Meghan Rubenstein, and Alexander Tokovinine. Learning goes both ways. In addition, this project has benefited greatly from the support of many people at The University of Texas at Austin, including Paola Bueché, John Yancey, Ken Hale, Doug Dempster, and Steve Leslie. Most of the book was written over the summer of 2009 at UT's beautiful center in Antigua, Guatemala, the Casa Herrera, and I appreciate their help in making all that happen.

My editor, Trace Murphy, first contacted me many years ago about the idea of such a book, and without his good advice and patient editing it would never have been written. And several people were of great help in providing illustrations, including Ricky Lopez Bruni, Jorge Perez de Lara, and Peter Mathews.

At last I give inadequate thanks to my family. George and Gene Stuart, my parents who have lived a life of archaeology, took me to Mexico and Guatemala when I was just three years old, and made everything possible. My elder siblings Ann, Roberto, and George, grew up in the same incredible environment of jungles and ruins, so they know also just how special it all was. And my sons Peter and Richard, the joys of my life, always inspire me today, even as they wait patiently for me to get away from the computer. To all of my family, and now my dear Carolyn, I offer thanks for giving order to my days.

INDEX

Note: page numbers in *italics* indicate illustrations.

agriculture, 31; calendar and, 104, 107;
    replanting metaphor, 269–71. *See
    also* maize
Ah Napot Xiu, 261
Aj Chak Wayib, 264
Aj Chan, 9, 16
Aj K'ulel, 69–70
alawtuns, 231, 232, 237
Alvarado, Pedro de, 122, 161
animate landscape, 68–69, 90
animate time, 152; time periods linked
    with rulers or gods, 20, 152,
    156–57, 183, 256–62, 268–74,
    277–78
animism, 67–70
Antiquities of Mexico (Kingsborough),
    286–87
Apocalypto, 284, 303
Argüelles, José, 306–7, 308
astronomy, 49, 99, 100, 101–2, 122. *See
    also* lunar cycles; Maya astronomy;
    planets; solar movements; stars
Avendaño y Loyola, Andrés de, 10–17,
    20, 257, 262

Aveni, Anthony, 192, 289
Aztec calendar, 113, 124–32, 145, 159,
    160; Calendar Stone, 90, 146,
    203–9. *See also* Aztec days
Aztec codices: astronomical records in,
    290, 292; calendrical content, 117,
    119, 120–24, *127*, 128–32, 202–3
Aztec days, 117–29; in the codices,
    119, 122–24, 202–3; day count in
    daily life, 125–27; day reckoning,
    115–16, 137; day signs' significance,
    119–22; divination and, 119–20,
    124–29; names and glyphs, 117–19,
    319–20; on the Calendar Stone,
    206–7, 208; year bearers, 129,
    131–32, 145
Aztec deities: in the codices, 122–23,
    124, 130–31, 202–3; Quetzalcoatl,
    41, 42, 63, 200–201, 307–8; sun
    god (Tonatiuh), 81, 120–22, 124,
    199–201, 206
Aztec mythology, 196–209; creation
    legends, 41, 196–203, 209, 305–6;
    on the Calendar Stone, 203–9

Printed in the United States
by Baker & Taylor Publisher Services